The Flow of Heat

The Flow of Heat

Keith Cornwell

Department of Mechanical Engineering
Heriot-Watt University, Edinburgh

VAN NOSTRAND REINHOLD COMPANY
New York - Cincinnati - Toronto - London - Melbourne

First published 1977
Reprinted 1978

**Published by Van Nostrand Reinhold Company Ltd.,
Molly Millars Lane, Wokingham, Berkshire, England**

*Published in 1977 by Van Nostrand Reinhold Company
A Division of Litton Educational Publishing, Inc.,
450 West 33rd Street, New York, N.Y. 10001, U.S.A.*

*Van Nostrand Reinhold Limited
1410 Birchmount Road, Scarborough, Ontario, M1P 2E7,
Canada*

*Van Nostrand Reinhold Australia Pty. Limited
17 Queen Street, Mitcham, Victoria 3132, Australia*

Library of Congress Cataloguing in Publication Data
Cornwell, Keith.
  The flow of heat.

  Includes index.
  1. Heat-Transmission.   I. Title.

TJ260.067   621.4'022   76-45658
ISBN 0 442 30177 4
ISBN 0 442 30168 5 pbk.

Printed in Great Britain by
Biddles Ltd., Guildford, Surrey

Production by Bucken Ltd.

To my Parents

Contents

PART II — AN INTRODUCTION TO THE ANALYSIS OF HEAT FLOW

Preface

In recent years there has been an increasing awareness that man is rapidly depleting the available energy resources. This awareness has not only initiated efforts to find new coal and oil reserves and use direct solar energy, but has also inspired attempts to reduce our energy consumption, both industrially and domestically. Most of the energy we use and most of the energy we waste is in the form of heat. In countries with a temperate or cold climate, a major proportion of the energy is used for heating human beings and there has recently been much publicity aimed at reducing domestic consumption. As a result of this energy-saving campaign and a general increase of interest in technical subjects, the ranks of those who desire some knowledge of heat and the flow of heat have swollen from the original group of thermal engineers and scientists. They now include the economy-minded factory or office manager, the enthusiastic layman who insulates the attic of his home and the thrifty housewife who reduces the central heating load in the centre of the day.

This book on the flow of heat, or heat transfer as the subject is often called, is written for two categories of readers. These are the professional students of the subject and the technically inclined laymen or those of other technical disciplines who desire some knowledge of the topic. For the first category the book is a student text suitable for diploma and degree courses and covers all the material generally included up to engineering honours degree level. Part I is sufficient for students of lower level courses or for those who discontinue the subject in one of the earlier years of their degree course. Part II contains the essence of the book for those studying the subject to higher levels. The treatment is in SI units and is particularly sympathetic to the reader, with emphasis on important points and many worked examples. For the second category of reader Part I forms a readable introduction to the topic and requires only limited knowledge of physics and mathematics. Part II may be used as a reference for further knowledge of specific areas and for this purpose a summary of important heat transfer relationships is included. The flow of heat is not a subject which is only worthy of study as a means to an end. It has an intrinsic interest which I shall attempt to illustrate by application to everyday situations, by mention of historical aspects and, here and there, by a little wandering from the direct and well-worn path.

I am indebted to the authors of the many textbooks, reference books and technical papers that have been used in the preparation of the book. Name and date identification is given in the text and full location details are included in the reference list. I am much obliged to my colleagues at Heriot-Watt University and particularly to Dr. A. J. Addlesee for his many helpful comments. Finally I wish to thank Mrs. Anne Edward for typing and retyping the manuscript, Mrs. J. Burnett for preparing the drawings and my wife for tolerating many evenings of non-communication.

Keith Cornwell,
North Berwick,
May, 1976

Introduction

What is energy? We know certain things about it. We know for example that it can flow in various forms such as heat energy, electrical energy and mechanical work. It may also be stored in various forms such as strain energy in a compressed spring, internal energy in a hot body and chemical energy in a fuel. Furthermore, Einstein showed at the beginning of the twentieth century that it is interconvertible with mass itself; that is to say the whole physical world is really a manifestation of energy. The characteristics of the various forms of energy can be identified. We could say for example that heat energy flows due to a temperature difference or express the internal energy of a material in terms of the activity of its atoms, but this brings us no nearer to answering the question: What is energy?

The fact is that we do not really know the answer. Most scientific and technological subjects commence with an acceptance of the concept of energy and treat the various forms of energy and mass as the stuff of the universe. Questions regarding the basic nature and existence of energy are more appropriate to the fields of philosophy and religion. Science cannot give reasons for the existence of energy or the presence of the physical universe. We ourselves are part of this physical universe, part of the energy which we seek to understand and due to this it may be inherently impossible for us to comprehend the existence of energy. However, this daunting thought need not deter us from studying the various characteristics of energy. Man has largely progressed to his present state of civilization by gleaning knowledge about it.

The subject of this book concerns just one of the many manifestations of energy, that of heat. It is further restricted to a study of the rate of flow of heat, a topic often called heat transfer. We are not concerned here with the conversion of heat into other forms of energy, such as the conversion of heat to work, called thermodynamics or the conversion of heat to electricity, called thermoelectrics. It was mentioned earlier that heat is the form of energy which flows due to a temperature difference and the subject of heat transfer is therefore only applicable to systems which involve different temperatures.

A popular science textbook used in the latter part of the nineteenth century was entitled *Deschanel's Natural Philosophy* (1888). In Part 2 Professor Deschanel concisely summarized the methods of heat flow to surrounding air:

'The cooling of a hot body exposed to the air is effected partly by radiation and partly by conduction of heat from the surface of the body to the air in contact with it. The activity of the surface conduction is greatly quickened

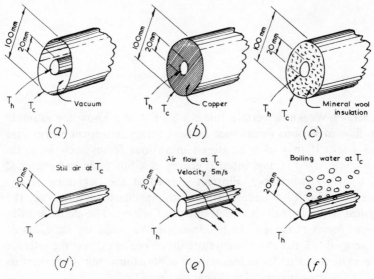

Figure 1   Heat transfer from a rod under various conditions.

	Predominant heat transfer mechanism	Dependence on temperatures	Heat transfer at steady state per m length with $T_c = 0\,°C, T_h = 100\,°C$
a	Radiation (blackbody)	$(T_h^4 - T_c^4)$	0.05 kW
b	Conduction	$(T_h - T_c)$	150 kW
c	Conduction	$(T_h - T_c)$	0.02 kW
d	Natural convection	$(T_h - T_c)^{1.25}$ approx.	0.07 kW
e	Forced convection	$(T_h - T_c)$ approx.	0.3 kW
f	Nucleate boiling	$(T_h - T_c)^3$ approx.	20 kW (with $T_c = 100\,°C, T_h = 120\,°C$)

by wind, which brings continually fresh portions of cold air into contact with the surface, in place of those which have been heated.'

We shall proceed to examine heat flow by these methods and in the same order, that is radiation, conduction and the activity of surface conduction or convection as it is called today. Furthermore, we shall not only consider the cooling of a hot body in air but also the heat flow from bodies in a vacuum, the heat flow through insulation, the heat flow from fluids in pipes and condensers and even the heat flow from our own human bodies.

Radiation and conduction are two basic and fundamentally different mechanisms. Radiation involves the propagation of energy through space at the speed of light and, like the transfer of light and radio waves, is an electromagnetic phenomenon. The situation in which radiation is the sole means of heat transfer only occurs when heat flows through a vacuum. On the other hand, conduction necessarily requires the existence of matter and depends upon molecular activity within the matter. Thermal convection is not a basic heat flow mechanism like radiation or conduction and mainly involves conduction together with the physical movement of matter. Professor Deschanel correctly described it as the activity of surface conduction, because the heat flow from a surface to a fluid (gas or liquid) is governed by the conduction process through a thin layer of the fluid adjacent to the surface. Convection heat transfer is subdivided into forced convection, where the fluid is forced over the surface by a blower or pump, and natural convection, where the temperature of the surface itself causes the convection currents. Convection mechanisms involving phase changes lead to the important fields of boiling and condensation. A rough indication of the magnitude of heat flow under various conditions is given in Fig. 1 and the accompanying table.

The Fundamentals
of Heat Flow

Heat Flow Through a Vacuum

1

1.1 Vibrations and Voids

The awareness of heat radiation has been with us ever since our primeval ancestors delighted in the rays of the sun or in the cosiness of their cave fires. However, it was not until the beginning of the twentieth century that any understanding of its nature was attained. Radiant heat energy was originally considered to propagate from a heat source rather like the ripples on a pond propagating from a point of disturbance. It could be shown that these ripples transferred energy by allowing the vertical oscillation of a float to operate a ratchet mechanism and raise a weight. In a similar way, it was assumed that radiation waves also transferred energy owing to their vibratory motion. Unfortunately, this did not account for the most significant aspect of radiation; its ability to pass through a vacuum. This ability was originally shown by Count Rumford using a thermometer situated in an evacuated glass chamber. Application of the wave theory requires some form of matter between the radiating bodies to enable transmission of the vibrations, or in other words there must be something to wave.

To circumvent this paradox, nineteenth century scientists postulated the existence of a gas called 'ether' which permeated all space (thus allowing for solar radiation) and penetrated the vacuum jars in their laboratories. Professor Deschanel in the aforementioned treatise writes:

'It is now generally admitted that both heat and light are due to a vibratory motion which is transmitted through space by means of a fluid called ether.' Perhaps the cautious manner in which he opens the sentence is significant. This was not the first time our forefathers had bridged the gap between their theories and experimental facts by postulating an invisible, weightless fluid with singular properties. Less than half of a century before, heat had been considered as a fluid named 'caloric' which flowed between bodies under conditions of 'heating by fire, rubbing or hammering'. When the possibility of converting heat into other forms of energy was demonstrated the concept of caloric was gradually abandoned although, as was mentioned earlier, energy is not really definable in general terms. Further progress with the wave theory of radiation was therefore severely hampered by the passage of radiation through a vacuum and, as we shall see later, it was not until the introduction of the quantum theory that the problem was resolved.

On another front, however, research was progressing satisfactorily, based on the concept of radiation as atomic particles. These particles formed a radiation gas, not a mysterious weightless gas, but one with characteristics and properties similar to a perfect gas. Radiation was known to pass through a vacuum at the velocity of light and, since this velocity is finite, there must be a finite amount of energy in transit at any time in the space between radiating surfaces. That is to say there must be a radiant energy density or radiation 'gas' occupying the space. By analysis of this gas along exactly similar lines to the classical kinetic theory of perfect gases Boltzmann showed in 1884 that the heat energy emitted by radiation from any surface is proportional to its absolute temperature to the fourth power:

$$E = CT^4$$

In this equation E is the quantity of energy emitted per unit area and per unit time by the surface and is termed the emissive power. The term T is the temperature, which is always measured in absolute units in radiation work, and C is a constant. This relationship confirmed the conclusions drawn by Stefan (1879), based on experiments conducted by Tyndall on radiation from hot platinum wire. Ironically Stefan's conclusions were rather inaccurate as it has since been shown that radiation from untarnished electrically conducting metals is more nearly proportional to T^5.

The emissive power of a surface is found to depend upon a number of parameters among which are the surface material and roughness. In order to formulate general radiation relationships, it is helpful to define a surface which is a perfect emitter and emits the maximum power possible at a particular temperature. We shall later show that a surface which is a perfect or ideal emitter may also be called a black surface. For a perfect emitter the previous expression becomes

$$E_b = \sigma T^4 \tag{1.1}$$

where E_b is the emissive power of the perfect emitter at temperature T. This equation is termed the Stefan–Boltzmann law and the Stefan–Boltzmann constant σ has the value of $56.7 \times 10^{-12} \, \text{kW/m}^2 \, \text{K}^4$. The emissive power of an actual surface is expressed as a proportion of E_b:

$$E = \varepsilon E_b \tag{1.2}$$

The proportion ε is termed the emissivity of the surface and its value depends upon surface characteristics and temperature.

1.2 Symbols and Surfaces

Radiation involves the use of many technical terms and definitions and before proceeding further we shall introduce the more important of these terms. The total radiant energy falling on a surface is called the incident radiation and the incident radiation per unit time and area is termed the irradiation H. There are three possibilities open to radiation striking a body; it may be absorbed, reflected or transmitted. The following parameters are accordingly defined:

Absorptivity α—the proportion of incident radiation absorbed

Reflectivity ρ—the proportion of incident radiation reflected

Transmissivity τ—the proportion of incident radiation transmitted

and it follows that

$$\alpha + \rho + \tau = 1 \tag{1.3}$$

Solids generally transmit no radiation unless the material is of very thin section. Metals absorb radiation within a fraction of a micrometre and electrical insulators within a fraction of a millimetre. Even substances such as liquids and glasses absorb most of the radiation within a millimetre. Solids and liquids therefore are generally assumed to have a transmissivity of zero in which case:

$$\alpha + \rho = 1 \tag{1.4}$$

On the other hand, most elementary gases such as hydrogen, oxygen and nitrogen (and mixtures of these such as air) have a transmissivity of practically unity; i.e., their reflectivity and absorptivity are nearly zero. For this reason radiation transfer through air is generally estimated using the relationships for radiation through a vacuum. Gases with a more complex structure, such as steam and carbon dioxide, generally absorb and emit as well as transmit radiation.

The reflection of radiation from a solid surface may be of a specular or diffuse nature. Specular reflection occurs at a surface which is very smooth, such as a mirror, and an image of the radiation source is projected. The optical laws apply and, in particular, the angle of reflection is equal to the angle of incidence. Diffuse reflection occurs when the surface is rough and there is no preferential direction of reflection. No actual body is perfectly specular or diffuse but it is often useful to approximate to one of these ideal surfaces.

Equation (1.3) indicates that when a body is such that no incident radiation is reflected or transmitted all the radiant energy must be absorbed. A body of this type is called a blackbody (and a surface which has this property is termed a black surface). For a blackbody therefore

$$\rho = 0, \ \tau = 0 \ \text{and} \ \alpha = 1$$

No actual body is perfectly black; if there were such a body it would not be possible to see it, except as a silhouette. Some surfaces are nearly black and it is possible to artificially create an almost perfectly black area by forming a cavity in a material, as shown in Fig. 1.1. Radiation passing through the hole into the cavity is repeatedly absorbed and reflected at the cavity walls until it is all absorbed. An even better black surface is provided by distant surroundings such as the night sky, but in this case one is restricted to the particular

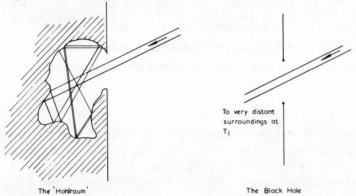

The 'Hohlraum'　　　　　　　　　The Black Hole

Figure 1.1　The black 'surface'.

temperature of the surroundings. It should be pointed out that surfaces which are nearly black, for radiation purposes, are not necessarily black to visible light, because the visible light wavelength range is only a small part of the overall thermal radiation range. White paper, for example, is nearly radiation black with an absorptivity of 0.97.

Let us now turn our attention to radiation leaving a surface rather than radiation falling on a surface. The total radiation leaving a surface per unit time and area is termed the radiosity, B. The radiation leaving a surface may be divided into two parts: that which is reflected by the surface, and that which is emitted by the surface due to its temperature. The former part is equal to ρH where H, it will be recalled, is the irradiation or the total incident radiation per unit time and area. The latter part is the emissive power E which was introduced earlier as the energy emitted per unit time and area. Substituting $E = \varepsilon E_b$ from equation (1.2) where E_b is the maximum possible emissive power, the radiosity becomes

$$B = \rho H + \varepsilon E_b \tag{1.5}$$

A little thought now shows that a blackbody is also a perfect emitter. Consider a blackbody in thermal equilibrium with surroundings at temperature T_a as shown in Fig. 1.2a. The incident radiation is equal to the total radiation leaving the body, therefore

$$H_1 = B_1$$

and since $\rho = 0$ for a blackbody equation (1.5) becomes

$$B_1 = \varepsilon E_b$$

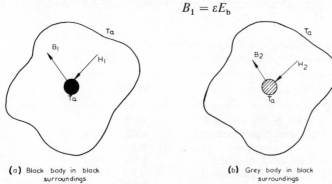

(a) Black body in black surroundings   **(b)** Grey body in black surroundings

Figure 1.2   Radiation to bodies in thermal equilibrium.

Furthermore, since by definition the blackbody absorbs the maximum amount of incident radiation possible from the surroundings at T_a (that is H_1), it also emits the maximum energy possible (that is E_b) to the surroundings at T_a to maintain equilibrium:

$$H_1 = E_b$$

From these equations it is evident that the emissivity is unity and for any blackbody therefore

$$\alpha = 1 \text{ and } \varepsilon = 1$$

Similar consideration of the case of a non-blackbody in thermal surroundings at a temperature T_a, as shown in Fig. 1.2b, gives under equilibrium conditions

$$B_2 = H_2$$

and from equation (1.5)

$$B_2 = \rho H_2 + \varepsilon E_b$$
$$= (1 - \alpha)H_2 + \varepsilon E_b$$

Rearrangement then yields

$$H_2 = \left(\frac{\varepsilon}{\alpha}\right) E_b$$

Since the properties of the body do not affect the radiation from the surroundings $H_2 = H_1$ and hence

$$\varepsilon = \alpha$$

The emissivity and absorptivity of a surface are therefore equal at any

7

particular temperature. A further limitation should be placed on this identity and the surfaces specified as radiation grey or, alternatively, the identity should be applied at a particular wavelength, but these aspects will be considered in Section 1.3. The identity was originally embodied in Kirchhoff's radiation laws of 1859. Table 1.1 gives emissivity (or absorptivity) values for a few surfaces and more extensive data are available in heat transfer reference texts.

TABLE 1.1 The approximate emissivity of various surfaces at ambient temperature

Surface	*Emissivity*
Brass, highly polished	0.03
Copper, highly polished	0.03
oxidized	0.8
Steel, mild, polished	0.07
galvanized	0.3
rusted	0.8
Brick, fireclay	0.7
red, building	0.9
Glass, polished	0.94
Paper, white	0.97
Paint, white-gloss	0.90
black-gloss	0.90
black-matt	0.97
aluminium	0.3–0.6
Water	0.95

The reader has now been introduced to sufficient notation to enable estimation to be made of the heat transfer by radiation from a body at one temperature to surroundings at another temperature. If the body shown in Fig. 1.2b is not in thermal equilibrium with its surroundings but is at a higher temperature, the heat received by the body is less than that dissipated, that is to say B is more than H. The net heat flow rate Q from the body of area A to the surroundings is given by

$$\frac{Q}{A} = B - H \tag{1.6}$$

Consider the situation shown in Fig. 1.3 where heat flows from the body at temperature T_1 to the surroundings at temperature T_2. Due to the fact that the surroundings are black, it follows that

$$H = E_{b2}$$

where suffix 2 denotes the temperature of the surroundings. From equation (1.5)

$$B = \rho_1 H + \varepsilon_1 E_{b1}$$

where suffix 1 denotes the temperature of the body.

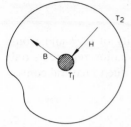

Figure 1.3   Grey body in black surroundings.

Noting that $\rho_1 = 1 - \alpha_1 = 1 - \varepsilon_1$, substitution in equation (1.6) gives

$$\frac{Q}{A_1} = \varepsilon_1 E_{b1} - \varepsilon_1 E_{b2}$$

and by using equation (1.1)

$$Q = -\sigma\varepsilon_1 A_1 (T_2^4 - T_1^4) \tag{1.7}$$

The negative sign is retained in this text so that the direction of heat flow in radiation is consistent with that in conduction ($Q = -kA(T_2 - T_1)/x$) and convection ($Q = -hA(T_2 - T_1)$). In each case Q is taken to mean the heat flow from body or surface 1 to 2 and positive heat flow therefore occurs down the temperature gradient.

As an illustration of the use of equation (1.7) let us investigate the possibility of reducing the heat loss from a domestic hot water tank by coating it with aluminium paint. The tank is 0.5 m diameter and 1 m high and is situated in a large space effectively forming black surroundings. (Distant non-black surroundings are effectively black because a negligible amount of energy is reradiated to the tank.) The estimation is based on a tank surface temperature of 80 °C and an ambient temperature of 25 °C. If the tank surface is oxidized copper with an emissivity of 0.8 the radiation heat flow is:

$$Q = -\sigma\varepsilon_1 A_1 (T_2^4 - T_1^4)$$
$$= -56.7 \times 10^{-12} \times 0.8 \times A_1 [(25 + 273)^4 - (80 + 273)^4]$$

or, in more convenient numerical terms:

$$Q = -56.7 \times 0.8 \times A_1 \left[\left(\frac{298}{1000}\right)^4 - \left(\frac{353}{1000}\right)^4 \right] \text{kW}$$

Substitution of $A = \pi/2 + 2\pi/16$ and evaluation gives

$$Q = 0.69 \text{ kW}$$

from the tank (1) to the surroundings (2). Some aluminium paints have an emissivity of about 0.3 and a coating of this paint on the tank would lead to a reduction in the heat loss of

$$\frac{0.8 - 0.3}{0.8} \times 0.69 = 0.43 \text{ kW}$$

It should be noted that white paint has an emissivity of 0.97 and would therefore increase the radiation loss. Heat loss by radiation represents only about half the total loss in this case. The other half is due to natural convection which is discussed in a later chapter.

1.3   Spectra and Quanta

The reader is now requested to consider once again the basic nature of radiation, particularly with regard to its wave-like form. It may be recalled that we mentioned the similarity between radiation and light. General observation has familiarized us with the concept that when a body is heated to high temperatures its colour changes from dull red to white. More precisely, as the temperature of a body is varied, not only the quantity of radiation emitted changes but also its distribution over the wavelength band. Towards the end of the nineteenth century a number of research workers studied this effect experimentally, and obtained curves for the distribution of radiation at various temperatures across the wave spectrum, which were similar in form to those shown in Fig. 1.4.

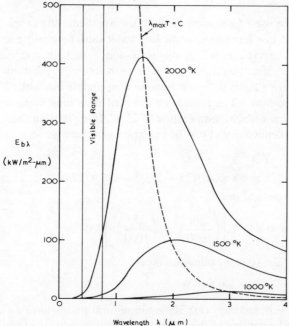

Figure 1.4   Spectral distribution of monochromatic emissive power.

W. Wien attempted to analyse the spectral distribution of radiation theoretically and published his conclusions, which subsequently became known as Wien's laws, in 1893 and 1896. By a classical thermodynamic and statistical analysis of the ideal radiation gas he derived a distribution law which may be expressed in terms of the single wavelength (or monochromatic) blackbody emissive power $E_{b\lambda}$ as

$$E_{b\lambda} = \frac{C_1}{\lambda^5} \left[\frac{1}{\exp(C_2/\lambda T)} \right] \tag{1.8}$$

where λ is the wavelength and C_1 and C_2 are constants. The wavelength λ_{max} giving the maximum emissive power at a particular temperature may be found by differentiating equation (1.8) and equating to zero:

$$\frac{dE_{b\lambda}}{d\lambda} = - \frac{5C_1}{\lambda^6 \exp(C_2/\lambda T)} + \left[\frac{C_1}{\lambda^5 \exp(C_2/\lambda T)} \right] \left[\frac{C_2}{\lambda^2 T} \right] = 0$$

The result is

$$\lambda_{max} T = \frac{C_2}{5} = C$$

where C is another constant. This expression is one of Wien's laws and has been verified experimentally.

Wien's distribution does not satisfactorily account for the emissive power–wavelength relationship at high temperatures. From the formula it can be seen that as T tends to infinity, $E_{b\lambda}$ tends to the finite value of C_1/λ^5 and this is not the case in practice. At the turn of the century Max Planck studied this discrepancy, together with another unsatisfactory distribution law derived from classical statistics by Lord Rayleigh, and came to the conclusion that the failing lay in the theory of classical statistics itself. This led him to develop the quantum theory which successfully explained these inconsistencies and, furthermore, eventually provided the link between the particle and wave-like nature of radiation which had so puzzled the nineteenth century scientists.

The quantum theory proposes that energy is not infinitely divisible but consists of discrete parts or quanta. Each quantum has an energy equal to the product of the frequency, v, of the energy wave and a constant h which has become known as Planck's constant. Thus

$$\text{Energy} = hv$$

Radiation, like any other form of electromagnetic wave, travels at the velocity of light c and so the wavelength may be expressed as

$$\lambda = \frac{c}{v}$$

By analysing the spectral distribution of quanta according to the laws of probability, Planck (1901) developed the following expression for the mono-

11

chromatic emissive power of a blackbody in a vacuum

$$E_{b\lambda} = \frac{2\pi hc^2}{\lambda^5}\left[\frac{1}{\exp(hc/\lambda K_0 T) - 1}\right] \tag{1.9}$$

where K_0 is Boltzmann's constant. The values of the constants are given in Table 1.2 and the variation of $E_{b\lambda}$ with λ is shown in Fig. 1.4. Differentiating equation (1.9) and equating to zero also yields Wien's law, $\lambda_{max}T = C$, and the value of the constant is included in Table 1.2. It is interesting to note that

TABLE 1.2   Radiation constants in SI units

Planck's constant	h	6.6256×10^{-37} kJ s
Boltzmann's constant	K_0	1.3805×10^{-26} kJ/K
Velocity of light	c	2.9979×10^8 m/s
Stefan–Boltzmann constant	σ	56.697×10^{-12} kW/m² K⁴
Wien's law constant	C	2.8978×10^{-3} m K

equation (1.9) differs from Wien's distribution, equation (1.8), only in the inclusion of the term -1 in the denominator and yet this step required the application of a completely new concept of radiation, a concept which eventually became the basis of modern physics.

The quantum theory, although proposed initially to explain anomalies in the classical theory of radiation, quickly spread to other forms of electromagnetic propagation. The final link in the fundamental physics of radiation was supplied in 1905 by Einstein's well-known relationship between energy and mass m; Energy $= mc^2$. We can therefore write:

$$mc^2 = h\nu = \frac{hc}{\lambda}$$

and

$$m = \frac{h}{c\lambda}$$

Thus the radiant energy quantum of wavelength λ may be considered as a particle of mass $h/c\lambda$. Particles of this type are identical to the 'particles' of the radiation gas which were analysed by Boltzmann and led to the Stefan–Boltzmann law. It also successfully explains how waves can travel through empty space without the necessity for a mysterious 'ether' because radiation through a vacuum may be considered as discrete quanta of energy or 'photons' propagating space at the velocity of light. Owing to the equivalence of energy and matter at the velocity of light these quanta may also be considered as particles of a radiation or photon gas.

Another piece of the radiation jigsaw which fitted neatly into place within the structure of the quantum theory was radiation pressure. As early as 1865 James Clark Maxwell had determined the effect of radiation pressure due to the impact of radiation gas particles on a surface. In terms of the quantum theory, this radiation pressure is equal to the rate of change of momentum of photons at a surface per unit area, and the momentum is given by mc or h/v. Recently, solar radiation pressure has become important in space technology as it has a measurable effect upon spaceships and the orbits of satellites.

The importance of the wavelength aspect of radiation may, however, be illustrated by the more down-to-earth example of a garden greenhouse. Window glass transmits radiation in the range of wavelengths from about 0.15 to 3 μm as shown in Fig. 1.5. It is almost opaque to radiation of longer wavelengths; i.e., longer wavelengths are absorbed or reflected. Most of the radiation which reaches the earth from the sun is within this range and solar radiation is therefore passed through the glass to the topsoil in the greenhouse.

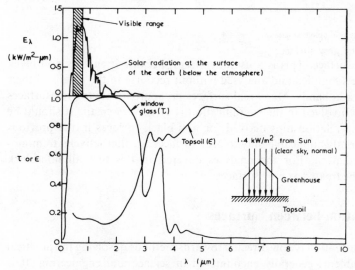

Figure 1.5 The greenhouse effect.

On the other hand, the topsoil radiates mainly in the longer wavelengths and reradiation from the topsoil to the surroundings is unable to pass through the glass. The heat transferred by solar radiation is therefore trapped in the greenhouse and causes the temperature to be higher than the external surroundings. The steady-state temperature in the greenhouse is dictated by an equilibrium between the radiant heat entering and the heat loss by various methods to the surroundings.

It was mentioned earlier that no actual surface radiates as a blackbody, although an artificial blackbody can be produced by forming a cavity. Furthermore, it is found that actual surfaces do not have a spectral emission which follows any smooth curve or distribution law. Nevertheless, it is convenient to define a surface which has a spectral distribution similar in form to that of a blackbody, but at a lower emissive power. The distribution shown in Fig. 1.6 has an emissive power which is a constant proportion of the blackbody emissive power at any wavelength; i.e., its emissivity at every wave-length is the same. A surface which has this type of spectral distribution is

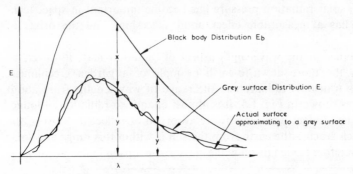

For a grey surface at all wavelengths $\dfrac{y}{x+y} \cdot \dfrac{E}{E_b} = \text{constant} = \epsilon$

Figure 1.6 The grey surface.

called a grey surface. This is a useful concept, because it enables non-black surfaces to be simplified and analysed without recourse to the details of their spectral emission bands. Many real surfaces do approximate to grey surfaces and the error involved in the simplification is often very small. It should be noted that the relationships derived for non-black surfaces in the previous section are strictly applicable only to grey surfaces or, alternatively, to mono-chromatic radiation. For the analyses covered in this text all non-black surfaces will be treated as grey surfaces.

1.4   Radiation between Surfaces

The reader's attention is now directed towards methods of solving the practical radiation problems generally encountered in science and engineering. It is perhaps at this point that the paths of the pure physicist and the applied scientist or engineer diverge; the physicist progressing with a deeper study of the nature of radiation and the applied scientist or engineer proceeding along the path which we shall follow.

The most common practical problem involves the estimation of heat transfer by radiation between two bodies. So far, consideration has only been given to the transfer between a body and the surroundings and, as the sur-roundings entirely enclose the body, the radiation involved has been the total quantity or the radiation in all directions from the body. In the case of radiation between two bodies only a proportion of the total radiation from each body reaches the other body and it is this proportion which is the concern of this section. Initially the exchange between blackbodies will be considered and then the treatment will be extended to cover greybodies.

Radiation between a blackbody and black surroundings may be expressed from equation (1.7) (with $\varepsilon = 1$) in the form

$$Q = -A_1\sigma(T_s^4 - T_1^4)$$

where suffixes 1 and s refer to the body and the surroundings. In the case of radiation to a part of the surroundings occupied by a second body, suffix 2, the net heat transfer to this body, Q_{12}, may be considered as a fraction F of the total heat transfer to the surroundings. That is to say:

$$Q_{12} = -A_1 F_{12}\sigma(T_2^4 - T_1^4) \tag{1.10}$$

There is no lack of variety in the literature for terms used to identify the factor F which may be called the shape factor, geometric factor, view factor, aspect ratio or configuration factor. A general analytical expression for the shape factor can be derived by consideration of the radiation intensity from a surface.

The radiation heat transfer between two small, parallel areas dA_1 and dA_2, as shown in Fig. 1.7, is found to be dependent not only on the areas but also on their distance apart. More precisely it is inversely proportional to the

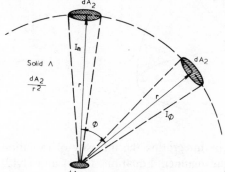

Figure 1.7 Radiation intensity.

square of the distance between them. (This is made readily apparent by using the analogy between radiation and light. The light per unit area falling on a screen from a slide projector, for example, is reduced by a factor of 1/4 when the distance between the screen and projector is doubled.) The following proportionality may be written for the heat flow between two incremental areas spaced a distance r apart

$$dQ \propto dA_1, dA_2, \frac{1}{r^2}$$

The constant of proportionality is the radiation intensity and, when the two areas are parallel, it is called the normal intensity I_n of radiation

$$I_n = \frac{dQ}{dA_1\, dA_2\, 1/r^2} \tag{1.11}$$

Alternatively, the normal intensity may be defined in terms of the solid angle as the radiation heat flow per unit area and per unit solid angle normal to the surface. The solid angle of surface dA_2 from dA_1 is given by dA_2/r^2 and substitution then leads to the same expression for the normal intensity.

The intensity I_ϕ at an angle to the surface other than the normal will depend upon the material and the surface geometry in the case of a real surface. The intensity versus angle characteristics of materials are discussed in the texts by Jakob (1949, 1957), Eckert and Drake (1972) and others. However, for a perfectly diffuse surface and, in particular, for a black surface it is simply dependent upon the angle ϕ to the normal;

$$I_\phi = I_n \cos \phi \qquad (1.12)$$

This expression is sometimes referred to as Lambert's cosine law.

The radiation passing from surface dA_1 to the surroundings may be considered as falling on a hemisphere of radius r as shown in Fig. 1.8. The total

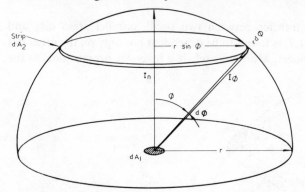

Figure 1.8 The black hemisphere.

radiation from dA_1 may be found by integrating the intensity of radiation over the hemisphere in the following manner. Equations (1.11) and (1.12) may be combined to give:

$$\frac{dQ}{dA_1} = I_n \cos \phi \, \frac{dA_2}{r^2} \qquad (1.13)$$

In this case dA_2 is the annular strip shown in Fig. 1.8 and substitution of its area yields

$$\frac{dQ}{dA_1} = I_n \cos \phi \times \frac{2\pi r \sin \phi r \, d\phi}{r^2}$$

$$= I_n \pi \sin 2\phi \, d\phi$$

Integration of the left-hand side gives the total heat flow per unit area from surface 1 which we have previously defined for the case when the surface is black as the total emissive power E_b. Thus:

$$E_b = \int_{\phi=0}^{\phi=\pi/2} I_n \pi \sin 2\phi \, d\phi$$

$$= I_n \pi \left[\frac{-\cos 2\phi}{2} \right]_0^{\pi/2}$$

$$= I_n \pi$$

and
$$I_n = \frac{E_b}{\pi} \tag{1.14}$$

We are now in a position to derive the relationship for the net radiation between two randomly orientated black incremental surfaces as shown in Fig. 1.9. In this case the right-hand surface as viewed from dA_1 is the pro-

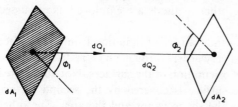

Figure 1.9 Radiation between black surfaces.

jected area of dA_2, that is $\cos \phi_2 \, dA_2$, and equation (1.13) for radiation from surface 1 to 2 becomes

$$\frac{dQ_1}{dA_1} = I_n \cos \phi_1 \, \frac{\cos \phi_2 \, dA_2}{r^2}$$

Substitution from equations (1.14) and (1.1) and rearrangement then gives

$$dQ_1 = \frac{\sigma T_1^4}{\pi r^2} \cos \phi_1 \cos \phi_2 \, dA_1 \, dA_2$$

Similarly for radiation from surface 2 to 1

$$dQ_2 = \frac{\sigma T_2^4}{\pi r^2} \cos \phi_2 \cos \phi_1 \, dA_2 \, dA_1$$

and the net exchange between the surfaces is

$$dQ_{12} = dQ_1 - dQ_2$$

or in integrated form;

$$Q_{12} = -\frac{\sigma(T_2^4 - T_1^4)}{\pi} \int_{A_1} \int_{A_2} \frac{\cos \phi_1 \cos \phi_2 \, dA_1 \, dA_2}{r^2} \tag{1.15}$$

Comparison of this equation with equation (1.10) yields the general expression for the shape factor and area product;

$$A_1 F_{12} = \frac{1}{\pi} \int_{A_1} \int_{A_2} \frac{\cos \phi_1 \cos \phi_2 \, dA_1 \, dA_2}{r^2} \tag{1.16}$$

17

Since the suffixes on the right-hand side of equation (1.16) are symmetrical, it follows that A_2F_{21} is equal to the same double integral and in particular

$$A_1F_{12} = A_2F_{21} \tag{1.17}$$

Thus, if the shape factor of surface 2 when viewed from surface 1 can be found, the shape factor of surface 1 from 2 can be determined readily from this relationship and this can lead to considerable simplification in many radiation problems. The calculation of A_1F_{12} or A_2F_{21} therefore involves the solution of a double area integral and in geometrically simple cases this may be possible mathematically. In the more complicated configurations generally encountered in practice other methods are employed as discussed in Chapter 6.

The reader has been shown that the transfer of radiation from one surface to another through a vacuum (and very nearly through air) is restricted in two ways. One restriction is due to the orientation of the surfaces. Not all the area available for radiation from one surface is covered by the second surface and this is taken into account by the shape factor and the area. The other restriction is due to the nature of the surface. Not all the radiant energy which could be emitted or absorbed is actually emitted or absorbed and this is taken into account by the surface properties. These restrictions to the flow of radiant energy may be considered as resistances, the former being a shape resistance and the latter a surface resistance. This concept enables an analogy to be made between radiation and the flow of current in an electrical network which can lead to considerable simplification of more complex systems.

The radiation network analogy involves considering the net heat flow rate between surfaces as a current and the differences between the radiosities of surfaces as potential differences. An expression for the shape resistance may be found as follows. The total radiation per unit area leaving surface 1 is the radiosity B_1 and the proportion of this energy which reaches a surface 2 is given by the shape factor:

$$\frac{Q_1}{A_1} = F_{12}B_1$$

Similarly for surface 2:

$$\frac{Q_2}{A_2} = F_{21}B_2$$

The net transfer of radiation from surface 1 to 2 is then

$$Q_{12} = A_1F_{12}B_1 - A_2F_{21}B_2$$
$$= A_1F_{12}(B_1 - B_2)$$

Comparison with Ohm's law relating current I and potential difference ΔV

$$I = \frac{1}{R}\Delta V$$

indicates that

$$\text{shape resistance} = \frac{1}{A_1 F_{12}} \tag{1.18}$$

Similarly an expression for the surface resistance may be found by considering the net radiation per unit area from a surface as

$$\frac{Q}{A} = B - H$$

Substitution of equation (1.5) with $\rho = 1 - \varepsilon$ yields

$$\frac{Q}{A} = B - \left(\frac{B - \varepsilon E_b}{1 - \varepsilon} \right)$$

that is

$$Q = \left(\frac{\varepsilon A}{1 - \varepsilon} \right) (E_b - B)$$

and comparison with Ohm's law indicates that

$$\text{surface resistance} = \frac{1 - \varepsilon}{\varepsilon A} \tag{1.19}$$

Two surfaces that 'view' each other may be considered as having three separate resistances to radiation transfer between them, as each surface has a surface resistance and there is a shape resistance due to their orientation in space and distance apart. The situation for two parallel grey surfaces of infinite extent is shown in Fig. 1.10 and a knowledge of the relevant parameters allows calculation of the heat flow, analogous to the current flow, between the surfaces. A study of this situation and more complicated systems involving a number of surfaces is undertaken in Chapter 6.

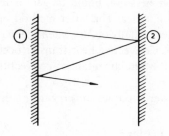

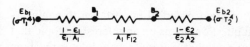

Figure 1.10 Radiation between infinite parallel grey surfaces.

Heat Flow through Matter

2

There are two basic methods of transferring heat using matter as the transfer medium. One method is to pass heat through the matter in a continuous manner from a high temperature zone to a low temperature zone. If, for example, one end of a metal bar is kept hot and the other end is kept cool heat will flow continuously through the bar. The other method is to physically move matter which is at a high temperature to a cooler region. An elementary physics experiment involves heating a block of copper, transferring it to a beaker of water and calculating the energy which has been transferred to the water. The former manner of heat transfer involving the continuous flow of heat through matter (whether that matter be solid, liquid or gas) is termed conduction and is the main topic of this chapter. The latter method of heat transfer, involving the movement of matter, usually occurs when the matter is a liquid or gas and is termed convection. Convective heat transfer occurs in many thermal engineering systems and the fundamentals of convection are considered in the following chapter.

In this chapter we shall attempt to answer two questions regarding thermal conduction:

How much heat is transferred through matter?
How is heat transferred through matter?

The first question is obviously of great practical importance to engineers and applied scientists. The second question opens up an interesting topic that is becoming increasingly important now that material technology is reaching a point where, within limits, desired properties can be designed into the material.

2.1 Thermal Conductivity

Let us start this investigation in the way in which many of us in the past have commenced our most elementary investigations, with a small wooden brick. We shall assume that the brick is homogeneous and shall try to find the rate at which heat passes through it. For this purpose an experimental rig is designed to maintain one side of the brick at a high temperature and the opposite side at a low temperature. A possible arrangement is shown in Fig. 2.1 where heat is supplied electrically to the hot face of the block and removed by a flow of cooling water at the cool face. The temperature difference across the block may be measured by using thermometers inserted in the

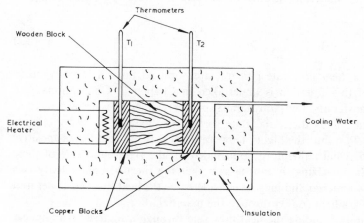

Figure 2.1 Experimental rig for measuring the thermal conductivity of the wooden block.

copper plates. Copper is used because, as we shall see later, it readily conducts heat and ensures that the temperature at the thermometer bulb is as near as possible to the temperature at the block face. The rate of heat flow through the block can either be found by measuring the power input of the electrical heater or by measuring the increase in temperature of the cooling water and its flow rate.

It would be found from careful measurements using our simple rig that the heat flow rate Q (in watts) was very nearly proportional to the temperature difference $(T_1 - T_2)$ across the block. It would not be quite proportional owing to deficiencies in the rig such as heat loss through the insulation, but neglecting these deficiencies for the moment the results of the experiment could be expressed as

$$Q \propto (T_1 - T_2)$$

In addition, the heat transfer is dependent upon the size of the block. The heat flow is increased if the area A of flow is increased and decreased if the distance x through which it flows is increased so that

$$Q \propto \frac{A}{x}$$

This could be checked experimentally by inserting in the rig and testing blocks of various configurations. Combination of these relationships leads to the equation of heat conduction in one direction which is called the Fourier equation:

$$Q = -k \frac{A}{x}(T_2 - T_1) \tag{2.1}$$

The constant of proportionality k is called the thermal conductivity and is a property of the material. In differential form the Fourier equation may be expressed as

$$q = -k \frac{dT}{dx} \tag{2.2}$$

where q is the heat flow rate per unit area or heat flux. A more rigorous derivation of this equation is undertaken in Chapter 7. Conventionally in both heat transfer and thermodynamics, property differences are expressed as the final state minus the initial state so that in this case we use the lower temperature T_2 minus the higher temperature T_1. If the heat flow is positive a negative sign must therefore be included on the right-hand side of these equations. Summarizing, it can be said that the thermal conductivity is a property of a material and may be considered as the heat flux (Q/A) per unit temperature gradient ($\Delta T/x$) through the material.

This method of finding the heat flow rate through a material may appear fairly obvious and straightforward to us today but it was not the method which occurred to scientists of the late eighteenth century. This is not surprising because the concepts of temperature and heat were vague and methods of measuring temperature were crude. Ingenhousz (1789) reported an interesting experiment designed to compare the heat flow rates through various solids. The apparatus is shown in Fig. 2.2 and consisted essentially of a metal box to which rods of various materials were attached. The procedure involved coating the rods with wax and filling the metal box with hot water.

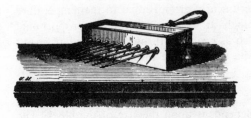

Figure 2.2 Apparatus used by Ingenhousz, 1789.
(From a wood engraving in Deschanel, 1888.)

Heat flowed from the metal box along the rods and melted the wax for a distance along each rod which depended on the material of the rod. In this way a comparison of the ability of various materials to conduct heat was made.

The concept of thermal conductivity emerged at the beginning of the nineteenth century with the studies of J. B. Biot (1804) and the publication of *Théorie Analytique de la Chaleur* by Joseph Fourier (1822). This allowed quantitative study of the heat flow rate through materials. It is interesting to note that Isaac Newton appears to have appreciated the true notion of heat energy as a mode of motion and also that heat is conducted through a gas by vibration of the particles.

The range of thermal conductivity values is shown in Table 2.1. The terms of equation (2.1) have the following units; Q—kJ/s or kW, A—m^2, x—m, ΔT—K, where K indicates an increment of 1 degree K (which is of course

TABLE 2.1   The range of thermal conductivity values at ambient temperature

Thermal conductivity k(W/m K)		*Thermal conductivity* k(W/m K)	
Solids		*Liquids*	
Copper	380	Water	0.61
Mild steel	54	Freon-12	0.071
Stainless steel (18/8)	16	Heat transfer oil	0.13
Concrete	1.4		
Wood (teak, across grain)	0.17	*Gases*	
Asbestos board	0.16	Hydrogen	0.18
Cork board	0.043	Air	0.026

equal to an increment of 1 degree C). Substitution therefore shows that the units of k are kW/m K. In this book conductivity values are generally multiplied by 10^3 and expressed in W/m K. Conversion factors to British units are given on p. 236. It is evident from Table 2.1 that the ratio of conductivity values between metals and good insulators is about 10^4. Although this may appear very large, it is considerably less than the ratio of some other properties and it causes heat loss to be an important factor in thermal systems. In contrast, the ratio of electrical conductivity between good and poor conductors is about 10^{24} and in electrical networks current loss through the insulation is therefore negligible.

2.2   Layers and Boundary Layers

In many practical situations involving heat transfer through materials the Fourier equation may be applied directly to yield the heat flow rate. The case of a multilayer wall, as detailed in Fig. 2.3, may be taken as an example.

23

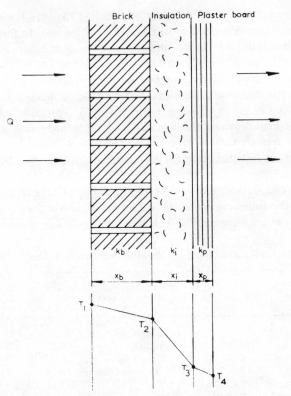

Figure 2.3 The composite wall.

Application of the Fourier equation (2.1) to the brick, insulation and plaster-board respectively yields:

$$Q = -\frac{k_b A(T_2 - T_1)}{x_b}; \; Q = -\frac{k_i A(T_3 - T_2)}{x_i}; \; Q = -\frac{k_p A(T_4 - T_3)}{x_p}$$

or $\quad T_2 - T_1 = -\frac{Q}{A}\left(\frac{x}{k}\right)_b; \; T_3 - T_2 = -\frac{Q}{A}\left(\frac{x}{k}\right)_i; \; T_4 - T_3 = -\frac{Q}{A}\left(\frac{x}{k}\right)_p$

Summation of the temperature differences gives

$$T_4 - T_1 = -\frac{Q}{A}\left[\left(\frac{x}{k}\right)_b + \left(\frac{x}{k}\right)_i + \left(\frac{x}{k}\right)_p\right]$$

or on rearrangement

$$\frac{Q}{A} = \frac{-1}{\left[\left(\frac{x}{k}\right)_b + \left(\frac{x}{k}\right)_i + \left(\frac{x}{k}\right)_p\right]}(T_4 - T_1) \qquad (2.3)$$

The term $(T_4 - T_1)$ is called the overall temperature difference, $\Delta T_{overall}$, and the purpose of heat transfer analysis is generally to determine the relation-

24

ship between the heat flux (Q/A) and the overall temperature difference. The expression which relates these two quantities is called the overall heat transfer coefficient U. In general therefore

$$Q = - UA \, \Delta T_{\text{overall}} \qquad (2.4)$$

and in the case of a multilayer wall the overall heat transfer coefficient based on the external wall temperatures is given by

$$U = \frac{1}{\left[\left(\dfrac{x}{k} \right)_b + \left(\dfrac{x}{k} \right)_i + \left(\dfrac{x}{k} \right)_p \right]}$$

If the wall had consisted of the brick layer alone, U would be given by k_b/x_b and equation (2.4) would simplify to the Fourier equation. (The negative sign in equation (2.4) is often omitted as the direction of heat flow is fairly obvious.)

From this analysis the reader may correctly infer that it is unnecessary to know the temperatures of the intermediate surfaces of a composite wall in order to deduce the heat flow rate; only the outer temperatures, T_1 and T_4, are required. Furthermore, in many instances not even these temperatures are known but rather the bulk temperatures of the fluids (either gases or liquids) which surround the wall. For example, in the calculation of heat loss from a domestic hot water tank the water temperature and the surrounding air temperature are more likely to be known than the temperature each side of the tank wall. Within the tank the water temperature is almost uniform owing to natural circulation and convection, but in a thin layer at the edge called the film or boundary layer the temperature drops as indicated in Fig. 2.4. Within this layer circulation is hampered by the viscosity of the fluid and heat is

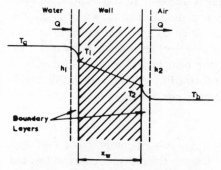

Figure 2.4 Temperature distribution at the tank wall.

mainly transferred by conduction. In a similar way another boundary layer exists on the air side. These layers act as insulation on the surfaces and further restrict the heat flow. This resistance to heat flow is taken into account by a parameter called the film coefficient, surface heat transfer coefficient or simply the heat transfer coefficient, h. The value of h under various conditions

25

will be extensively discussed in the following chapter and for the moment the reader is requested to accept it as indicating the surface conductance. The heat transfer coefficient is defined by the following equation (which in a slightly different form is attributable to Isaac Newton):

$$Q = -hA(T_2 - T_1) \qquad (2.5)$$

The temperatures in this case refer to the bulk fluid and the surface (and for heat flow in a positive direction T_2 is the lower temperature, whether that be the bulk fluid or the surface). The units of h are kW/m^2 K, or in the British system, Btu/ft^2 h °F.

In the example of the hot water tank wall shown in Fig. 2.4 the heat flow equations for the inner surface, the wall and the outer surface are respectively:

$$Q = -h_1 A(T_1 - T_a)$$

$$Q = -\frac{k_w}{x_w} A(T_2 - T_1)$$

$$Q = -h_2 A(T_b - T_2)$$

Rearrangement and summation as in the case of the composite wall leads to

$$\frac{Q}{A} = -\left[\frac{1}{\dfrac{1}{h_1} + \left(\dfrac{x}{k}\right)_w + \dfrac{1}{h_2}} \right] (T_b - T_a)$$

and comparison with equation (2.4) shows that the term in the first bracket on the right-hand side is the overall heat transfer coefficient U.

It has been assumed that heat transfer is due only to convection and conduction. This is not generally the case as radiation can play a considerable part in the heat transfer from surfaces even at low temperature, as was discovered in the previous chapter. In the case of a domestic hot water tank it may well account for about half of the heat loss. The total surface heat transfer coefficient h_t may therefore be formed by adding a radiative coefficient h_r to the convective coefficient h already considered so that

$$h_t = h + h_r$$

The case of a wall between two fluids which is itself composed of a number of layers may be analysed along similar lines to the previous cases and would lead to the result indicated in Fig. 2.5. In general the analysis of plane walls under steady-state conditions may be summarized as:

$$Q = -UA \, \Delta T_{overall}$$

$$\frac{1}{U} = \sum \frac{1}{h} + \sum \frac{x}{k} \qquad (2.6)$$

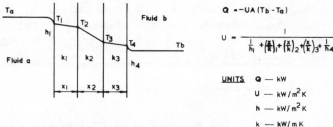

Figure 2.5 The overall heat transfer coefficient U.

2.3 Heat Flow through Solids

The reader's attention is now directed towards a consideration of the manner in which heat energy flows through the agglomeration of atoms and molecules which constitute matter. In this section we shall restrict our study to solids and in the following section we shall progress to the liquid and gaseous phases.

In solids there are two distinct mechanisms for transferring thermal energy. One mechanism is due to vibrations which are transferred by the structure or lattice of the solid and the other is due to free electrons which move about within the lattice and, in addition to carrying an electrical charge, also transfer heat. In the case of metals at normal temperatures the quantity of heat transferred by this second mechanism is, typically, two orders of magnitude greater than that transferred by lattice vibrations. For this reason pure metals with a large number of free electrons have a considerably higher thermal conductivity than non-metals or dielectrics which have no free electrons. Metal alloys and semiconductors contain a limited number of free electrons and have comparable contributions to their conductivities from both lattice and electronic mechanisms. In general the conductivity of a solid material may be expressed as

$$k = k_{ph} + k_e \tag{2.7}$$

where k_{ph} and k_e denote the lattice and electronic components of the thermal conductivity. The suffix 'ph' is used as lattice vibrations are more correctly called phonons. Phonons are standing waves which travel through material at the speed of sound and are not to be confused with photons which travel at the speed of light. These two conduction mechanisms will now be considered in detail.

Solid materials which do not conduct electricity transfer heat energy entirely by lattice vibrations or phonons. If the phonons are treated as particles in a similar way to the photons of radiation in the previous chapter, it can be shown by use of the classical kinetic theory (see, for example, Kittel, 1956) that the thermal conductivity is given by the following expression:

$$k_{ph} = \tfrac{1}{3}\rho c_v v \lambda \tag{2.8}$$

In this expression ρ represents the density of the material, c_v the specific heat at constant volume, v the average velocity of the phonon 'particles', which is normally taken as the velocity of sound, and λ the mean free path of the particles between collisions. It is interesting to note that combination of this equation with the Fourier equation (2.1) yields the following relationship for the heat flux through a specimen of length l:

$$\frac{Q}{A} = \frac{1}{3}(\rho c_v \, \Delta T)\left(\frac{v\lambda}{l}\right)$$

This indicates that the heat flow rate is dependent on the product of the excess energy one end of the specimen with respect to the other end ($\rho c_v \, \Delta T$) and the effective transport velocity ($v\lambda/l$). The phonon mean free path is an important parameter in limiting the thermal conductivity of dielectrics. It indicates the degree of scattering of the lattice vibrations or, put another way, the attenuation of the sound waves.

In the hypothetical case of an ideal crystal of a pure material where every molecule occupies its geometrically correct position and there are no impurity molecules present, there would be no scattering of vibrations and no attenuation of the energy transfer rate through that crystal. That is to say the values of λ and k_{ph} would be infinite. Experimentally, very high values of k_{ph} have been measured in practically pure crystals at low temperatures, as indicated in Fig. 2.6. Sapphire at 50 °K, for example, has a thermal conductivity which

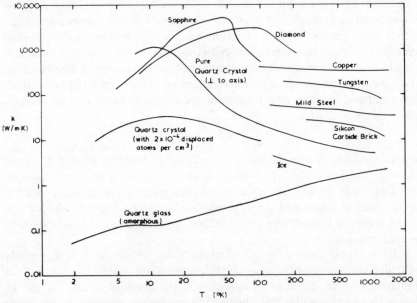

Figure 2.6 Variation of conductivity with temperature for some solids.

is an order of magnitude higher than copper at room temperature. But even this value of thermal conductivity is a long way from infinite and there is therefore some other limitation to heat transfer. Apart from the fact that the purest practical crystal has many imperfections, the value of k_{ph} is limited by the physical size of the crystal. The mean free path can under no condition be greater than the crystal dimensions. This size effect was first noticed experimentally by Haas and Biermasz (1935) and their results were successfully correlated by substituting the smallest linear dimension of the crystal for the value of λ.

In less pure crystals scattering due to impurities, interactions between phonons (such as Umklapp processes) and other effects lead to a much reduced mean free path. Each of the processes involved has been examined theoretically by Peierls (1955), Ziman (1960) and others and it has generally been found that λ is inversely proportional to the absolute temperature at ordinary temperatures and above. Since c_v and v are reasonably constant, this implies that k_{ph} is approximately proportional to $1/T$ and this has been verified experimentally. In ordinary crystalline materials consisting of a large number of small crystals in random orientations there is a further resistance to heat flow caused by scattering at crystal boundaries, dislocations, cracks and pores.

In the foregoing discussion of electrically non-conducting materials, only the crystalline substances have been considered. The amorphous or glassy materials have no long-range crystalline ordering of molecules although there may be local ordering in small clusters. In this respect they are similar to liquids and are often considered structurally as sub-cooled liquids. The general disorder and absence of a well-defined lattice leads to a low conductivity as indicated in Fig. 2.6. The conductivity of amorphous materials generally increases slightly with increased temperature.

Many electrically non-conducting materials commonly used in engineering are neither entirely crystalline nor entirely amorphous but consist of a mixture of these forms together with pores containing a liquid or air. Examples are refractory materials, wood, plastics, building laminates and thermal insulation. The thermal conductivity of these heterogeneous materials is really made up of contributions from each of the constituents together with the effects of the boundaries and convection in the pores. Quoted values of k are therefore apparent or overall thermal conductivities.

Let us now briefly consider the other component of heat conduction in solids; the electronic mechanism. Metals and semiconductors have a large number of free or valence electrons and these electrons conduct electricity and also transfer heat energy. In pure metals the electronic component of thermal conductivity is typically two orders of magnitude greater than the phonon component and in alloys and semiconductors there may be similar contributions from each. As both the electrical conductivity σ and the electronic part of the thermal conductivity k_e are dependent on the flow of electrons it is possible to derive a relationship between them. This relationship, termed the Wiedemann–Franz law is of the form

$$\frac{k_e}{\sigma} = LT \qquad (2.9)$$

29

where T is the absolute temperature and L is the Lorenz number of 2.45×10^{-8} $W\Omega/K^2$. Initially the correlation was solely based on metals at room temperature and it was L. Lorenz (1872) who indicated the proportionality with temperature. This relationship may be used to estimate the lattice conductivity in alloys and semiconductors where there are contributions from both mechanisms by combination with equation (2.7):

$$k_{ph} = k - \sigma LT$$

The interaction of the electron flow and the lattice vibrations can lead to minor errors, and an alternative method of separating the contributions from the two conduction mechanisms involves the use of a magnetic field aligned to reduce the electron flow by a known amount.

When a crystalline solid melts, its lattice framework disintegrates and the atoms or molecules become disordered. The increased scattering of lattice vibrations leads to a decrease in the value of k_{ph} on melting. The value of k_e also tends to drop on melting and the magnitude of this drop can be estimated from the variation in electrical conductivity. A liquid therefore generally has a thermal conductivity which is less than the solid, although bismuth is a notable exception (see Eckert and Drake, 1972). As liquid properties are so important in many thermal systems and processes their thermal conductivity is discussed in some detail in the following section.

2.4   Heat Flow through Liquids and Gases

The thermal properties of liquids are generally difficult to predict, which is unfortunate because they are also difficult to measure experimentally. The main problem in the measurement of thermal conductivity is caused by convection currents giving spuriously high heat flow rates. One method of overcoming this problem is to arrange for the liquid under test to be contained as an extremely thin layer so that convection currents are inhibited by the viscous shear forces. Another method involves passing a current through a wire immersed in the liquid and measuring the rate of heat flow and the rate of temperature increase with time. In this way the thermal conductivity can be measured in a second or two and before convection currents become noticeable. Whatever method is used, it is found that elaborate experimental precautions must be taken if reasonably accurate values of conductivity are to be obtained. Owing to these difficulties, there is a notable absence of data on the thermal conductivity of liquids at temperatures other than ambient, and some importance has been attached during recent years to predictive methods.

The general fluidity of liquids suggests that their molecular arrangement is rather like a compressed gas in which the general mobility of the molecules is slightly restricted but still random. On the other hand, consideration of the small change in value of some properties between the solid and liquid phase suggests that the structure of the liquid is more comparable to a solid than a gas. For example, a solid on melting typically increases in volume by a small fraction while a liquid on evaporation may increase in volume 1000 times. The increase in volume on melting is caused by the formation of vacancies in the lattice. These vacancies allow the molecules to have a limited mobility within the lattice, similar to the way in which the removal of one square in the arrangement shown in Fig. 2.7 allows a progressive movement of all the other

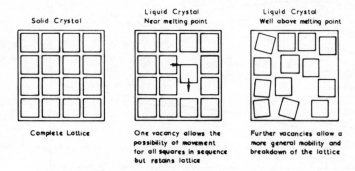

Figure 2.7 The liquid lattice model.

squares. The most important point this illustrates, however, is that the formation of a vacancy and the subsequent mobility does not destroy the lattice. X-ray diffraction studies on liquids have indeed shown that the molecules tend to cluster in groups having a weak lattice structure or short-range order as it is often called. These clusters are not structurally orientated to each other and there is no long-range order as in solids. From our discussion of solids, it is very evident that the lattice component of the thermal conductivity is sensitive to the structure, and the short-range order, therefore, has a strong influence on the conductivity of many liquids. When the temperature of the liquid is raised to values considerably in excess of the melting point, the expansion and formation of more vacancies leads to a further breakdown of the lattice and a further decrease in k_{ph}. For electrically non-conducting liquids, therefore, the value of k_{ph} generally decreases with temperature as is evident from Fig. 2.8.

Eventually, on still further increase of temperature, the liquid evaporates into the gaseous phase where there is no lattice and k_{ph} becomes zero. As we shall see later, gases do conduct heat to a small extent owing to the general mobility of the molecules.

The conductivity of liquids may be divided into a phonon and an electron component in a similar way to solids. More precisely the division should be between a phonon and translational component k_t because ions as well as

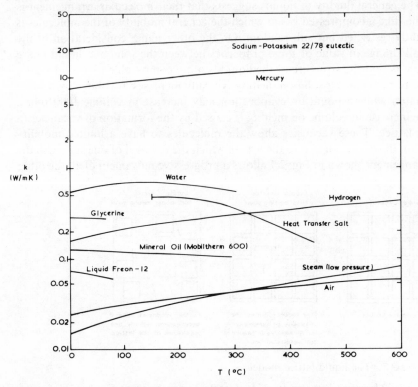

Figure 2.8 Variation of conductivity with temperature for some fluids.

electrons may transfer heat by movement or translation in a liquid:

$$k = k_{ph} + k_t \tag{2.10}$$

As in the case of solids, electrically non-conducting liquids are considered initially and are followed by a discussion of liquids in which translation plays a part.

Research into the mechanism of a natural phenomenon generally involves experimental work together with the formation of a 'model'. The model may initially be no more than an idea of the mechanism of the process in the mind or the notebook of the researcher. The model is then studied and analysed mathematically to yield relationships between parameters which can be measured experimentally. Discrepancies between the theoretical and experimental values may involve amendment or replacement of the model, recalculation of the theoretical values and further comparison with the experimental values in an iterative manner. A pattern of this type has developed from research into the thermal conductivity of liquids although the following simple analysis has led to very useful results in this case.

Consider a model in which a liquid consists of molecules arranged in a cubic array as shown in Fig. 2.9. Energy flows from the left to the right and

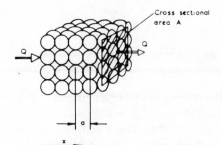

Figure 2.9 The cubic array of molecules.

the internal energy per unit mass, or specific internal energy u, decreases in the x direction. The change of specific internal energy between adjacent rows spaced a distance a apart is therefore given by $-a(\mathrm{d}u/\mathrm{d}x)$. The effective mass per row is given by $aA\rho$ where A is the flow area and ρ is the density, and the energy difference between adjacent rows is given by the product of these terms, $-a(\mathrm{d}u/\mathrm{d}x)aA\rho$. The velocity of energy transfer through the lattice is the phonon velocity or the velocity of sound v and the time taken to transfer an energy increment through a distance equal to the molecular spacing is therefore a/v. The energy flow per unit time or heat flow rate, Q, therefore becomes:

$$Q = \frac{-a(\mathrm{d}u/\mathrm{d}x)aA\rho}{a/v}$$

or in terms of heat flux;

$$q = -av\rho \frac{\mathrm{d}u}{\mathrm{d}x} \tag{2.11}$$

At constant volume $\mathrm{d}u$ is given by $c_v\mathrm{d}T$ where c_v is the specific heat at constant volume. The equation then becomes

$$q = -av\rho c_v \cdot \frac{\mathrm{d}T}{\mathrm{d}x}$$

and comparison with the Fourier equation (2.2) yields

$$k_{\mathrm{ph}} = \rho c_v v a \tag{2.12}$$

This expression is of a similar form to equation (2.8) for solid materials and this is to be expected as the basic mechanism in each case is the same (although λ and a are not directly comparable).

Equation (2.12) may be expressed in another way by making the following substitution where M denotes the molecular mass, N is the number of molecules per mole of material (or Avogadro's number), d is the molecular 'diameter', R_0 is the universal gas constant and K_0 is Boltzmann's constant, $(1.38 \times 10^{-26} \text{ kJ/K})$

$$\rho = \frac{M}{Nd^3} \tag{i}$$

$$c_v = \frac{3R_0}{M} \qquad \text{(ii)}$$

$$a = d \qquad \text{(iii)}$$

$$R_0 = K_0 N \qquad \text{(iv)}$$

The first expression arises by assuming the volume per mole is composed of N molecules each with a volume of d^3 and the second is taken directly from classical kinetic theory (see, for example, Kittel, 1956). The third follows from the packing geometry and the fourth is a fundamental relationship between physical constants. Substitution then leads to

$$k_{ph} = \frac{3K_0 v}{d^2} \qquad (2.13)$$

This relationship was originally derived by Bridgeman (1923) and yields values of conductivity which are surprisingly accurate in view of the assumptions made. Table 2.2 shows a comparison between the conductivity given by this model and experimental values for a few liquids. It correctly predicts that water has a value considerably greater than most other liquids. It is interesting to note that the thermal conductivity of water was first measured with some precision by Guthrie who reported his findings to the Royal Society in 1869.

TABLE 2.2   Comparison of computed and observed conductivities of some liquids at 30°C (from Jakob, 1949)

Liquid	Velocity of sound v (m/s)	Molecular distance d (m $\times 10^{10}$)	Thermal conductivity, k Theoretical (W/m K)	Observed (W/m K)
Water	1500	3.10	0.641	0.610
Ether	920	5.59	0.121	0.137
Acetone	1140	5.00	0.188	0.179
Methyl alcohol	1130	4.08	0.279	0.211

Many of the properties of water are anomalous and thermal conductivity was found to be no exception. As is shown in Fig. 2.8 the conductivity reaches a maximum value at about 130°C. The model we have studied has many limitations and the topic is obviously open to more refined and sophisticated treatments. These are described in texts on the thermal conductivity of liquids such as Tsederberg (1965).

Let us now briefly turn our attention to electrically conducting liquids such as metals and fused salts. Liquid metals and salts are used as high temperature heat transfer media in some nuclear power plant where they convey heat from the reactor core to the steam boiler. It was mentioned earlier that both liquid metals and fused salts conduct electricity and some heat energy by translation of electrons and ions respectively. (The term 'fused salts' is usually given to melted pure salts, such as sodium nitrate at temperatures above its melting point of 307 °C, or mixtures of pure salts above the eutectic temperature and does not mean aqueous electrolytes of salts in solution. Both fused salts and aqueous electrolytes transfer electricity by movement of ions.)

Liquid metals may be considered to have contributions to their thermal conductivity from both electrons and phonons in a similar way to solid alloys. Very often the electronic component is much greater than the phonon contribution and the thermal conductivity may therefore be estimated from the Wiedemann–Franz law (equation (2.9)). The conductivity of liquid metals is typically 100 times greater than that of other liquids. Fused salts have thermal conductivity values which are typically of the same order as that of water. The conductivity is mainly due to the phonon mechanism, as in electrically non-conducting liquids, although there is a small contribution from the diffusion of the ions. The phonon mechanism is largely dependent on the molecular mass of the ions and salts with a high mean ionic mass tend to have a low value of thermal conductivity (Cornwell, 1971). In heat transfer processes eutectic mixtures of salts rather than pure salts are often used in order to reduce the melting point. A mixture, commonly referred to as 'heat transfer salt' consists of 44% KNO_3, 49% $NaNO_2$ and 7% $NaNO_3$ and has a melting point of 142 °C.

In summary it can be said that most liquids fall into one of two categories; the phonon conductors and the electron conductors. In the former class ionic salts and water have conductivities of around 0.3 to 0.6 W/m K and other liquids have values which are typically around 0.1 to 0.2 W/m K. The latter class, consisting mainly of liquid metals, has conductivity values which are of a similar order to solid metals.

Gases differ considerably from liquids and solids and the major modes of heat transfer in processes involving gases are normally convection and radiation rather than conduction. Nevertheless the thermal conductivity of gases is important because to some extent it determines the convection characteristics. The conduction mechanism in gases is entirely different to that in liquids and solids. The molecules in liquids and solids are packed fairly tightly and there is some degree of order in their arrangement, while in gases each molecule may be separated from its neighbour by 100 or a 1000 times its diameter and its movement is entirely random. When the temperature of a gas is increased its internal energy is also increased. On the microscopic scale this internal energy is stored as the kinetic energy of the molecules and it may therefore be related to the mean velocity of the molecules. The thermal conductivity is determined by the rate at which energy can be transferred through the gas by collisions, and this in turn may also be related to the mean velocity of the molecules. From this rather crude picture of a gas it may therefore be correctly inferred that the conductivity increases as the temperature increases (as indicated in Fig. 2.8).

It would be wrong, however, to leave the reader with the impression that the thermal conductivity of gases can be predicted by an analysis of the random movements of the molecules. In fact an analysis along the lines of the classical kinetic theory (see for example Jakob, 1949) predicts a dependence of conductivity on the product of viscosity and specific heat, $k = \mu c_v$, and this relationship leads to values which are typically half those determined experimentally. In contrast to the theory of conductivity in some liquids this is a case where a simple model does not lead to a substantially correct prediction. A more rigorous treatment involving non-equilibrium aspects of heat flow, the form of the molecules and the dynamics of the collisions between the molecules must be used. The nature of the molecular collisions involves further models such as the Lennard–Jones 6–12 model which relates the distance between the molecules to the intermolecular forces. The prediction of gas thermal conductivity at very high temperatures has become important, as experimental determination is difficult owing to problems of containment and radiation. Values at these high temperatures are required in the development of gas turbines, rockets and in systems involving ionized gases or plasmas such as magnetohydrodynamic generators and, possibly in the future, nuclear fusion power reactors.

2.5 Radial Heat Flow

Metal tubes are involved in the essential parts of power stations, oil refineries and most process industries. The boilers have tubes in them, the condensers contain banks of tubes, the heat exchangers are tubular and all these items are connected by tubes. The radial heat transfer rate through a tube and any insulation which may surround it is therefore important and we shall consider the heat flow along similar lines to our study of plane walls in Section 2.2. Fig. 2.10a shows the general notation we shall use.

The temperature gradient through an annulus dr is given by $-dT/dr$ and the heat flow rate (from equation (2.2)) is therefore

$$Q = - kA \frac{dT}{dr}$$

Substitution of $2\pi r l$ for the area and integration between the inside, 1, and the outside, 2, yields

$$Q \int_{r_1}^{r_2} \frac{dr}{r} = -2\pi k l \int_{T_1}^{T_2} dT$$

$$Q = \frac{-2\pi k l (T_2 - T_1)}{\ln(r_2/r_1)} \qquad (2.15)$$

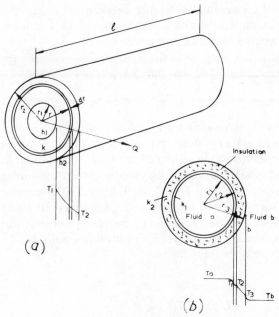

Figure 2.10 Radial heat flow through a tube.

It is more common in practice to know the temperatures of the fluids inside and outside the pipe rather than the pipe wall temperatures. This necessitates a knowledge of the surface heat transfer coefficients before the heat flow can be estimated (as in the previously discussed case of plane walls). In addition the pipe may have a layer of insulation or some other material around it as shown in Fig. 2.10b. The heat flow through each thermal resistance in this case is as follows:

$$Q = -h_1 2\pi r_1 l(T_1 - T_a) \qquad \text{from equation (2.5)}$$

$$Q = -\frac{2\pi k_1 l}{\ln(r_2/r_1)}(T_2 - T_1) \qquad \text{from equation (2.15)}$$

$$Q = -\frac{2\pi k_2 l}{\ln(r_3/r_2)}(T_3 - T_2) \qquad \text{from equation (2.15)}$$

$$Q = -h_3 2\pi r_3 l(T_b - T_3) \qquad \text{from equation (2.5)}$$

Thus $\quad T_b - T_a = (T_1 - T_a) + (T_2 - T_1) + (T_3 - T_2) + (T_b - T_3)$

$$= -\frac{Q}{l}\left[\frac{1}{2\pi r_1 h_1} + \frac{\ln(r_2/r_1)}{2\pi k_1} + \frac{\ln(r_3/r_2)}{2\pi k_2} + \frac{1}{2\pi r_3 h_3}\right]$$

or in general terms:

$$\frac{Q}{l} = -U'\Delta T_{\text{overall}} \qquad (2.16)$$

37

where $\Delta T_{overall}$ is the overall temperature difference between the fluids, Q/l is the heat flow per unit length of tube and U' is the overall heat transfer coefficient based on a unit length of tube. In generalized terms:

$$\frac{1}{U'} = \sum \frac{1}{2\pi h r} + \sum \frac{\ln(r_{out}/r_{in})}{2\pi k} \tag{2.17}$$

The heat flow per unit length rather than the heat flow per unit area is used in radial conduction because the area of flow obviously depends on the radius under consideration. This leads to an interesting effect in that the addition of insulation to a tube or rod of small diameter may under certain conditions increase rather than decrease the heat flow from the rod. Consider for example a rod of radius r_1 surrounded by insulation with a conductivity k and an outer radius of r_2. Heat is generated in the rod by the passage of electrical current and the rod surface is at a temperature of T_1 with surroundings at T_s. Equations (2.16) and (2.17) yield for this situation:

$$\frac{Q}{l} = -U'(T_s - T_1)$$

$$\frac{1}{U'} = \frac{1}{2\pi h_2 r_2} + \frac{\ln(r_2/r_1)}{2\pi k}$$

The radius of insulation which gives the maximum heat flow from the rod occurs when $dQ/dr_2 = 0$. Evaluation of the differential (which is left to the reader) shows that this radius is equal to the ratio k/h_2. If the radius of insulation is less than this value, the heat flow rate would be increased above the heat flow rate which would occur if there were no insulation. This occurs because, although increase in the thickness of insulation increases the thermal resistance due to the insulation, it also enlarges the area of the outer surface and therefore reduces the resistance due to the surface film. In practice the radius equal to k/h_2 is often small and less than the radius of the rod or tube, in which case there is no critical radius which yields a maximum heat flow rate.

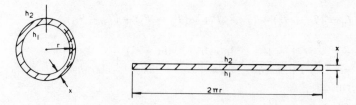

Figure 2.11 Approximation for a thin-walled tube.

When the tube wall is thin and has no insulation, radial heat flow through the tube may be simplified by using the mean radius and treating the problem as heat flow through a plate as indicated in Fig. 2.11. The heat flow equation then becomes:

$$Q = -UA\,\Delta T_{\text{overall}}$$

$$= -U2\pi rl\,\Delta T_{\text{overall}}$$

where

$$\frac{1}{U} = \frac{1}{h_1} + \frac{x}{k} + \frac{1}{h_2}$$

In the case of a tube of 25 mm bore with a wall thickness of 1 mm the error involved in the conduction term (x/k) by this approximation is about 1%. Often the conduction term is very much less than the $1/h$ terms and may be neglected entirely, as is the case in Example 5.3.

2.6   Heat Flow Prevention

Thermal insulation of the tubes we have been considering and of any engineering system is always a compromise between the prevention of heat flow and the minimization of the capital cost. It is possible to insulate a system almost perfectly by placing it in a vacuum and using multilayer aluminium foil techniques. In fact the loss of heat from a system at 100 °C to room temperature surroundings with a 100 mm thickness of this type of insulation would be about one-quarter of a watt per square metre. The cost of this insulation is generally prohibitive for normal engineering purposes excepting possibly cryogenic systems. The felt lining in the roof of a house, the thickness of insulation around a pipe and the padding or feathers in a sleeping bag are all compromises between thermal effectiveness and cost.

The materials which offer greatest resistance to heat flow by conduction are gases. But gases allow a considerable heat transfer by radiation and convection, so that the heat conveyed by conduction is only a small fraction of the total. However, it is possible to limit the convection currents and radiation transfer by containing the gas, which is usually air, in a fibrous or porous material. If the composite material of air and fibres has the major part of its volume occupied by air the effective conductivity of the material will be of a similar order to that of air. For example, when wool is packed at an apparent density (wool and air) of 80 kg/m^3 it has an effective conductivity (wool and air) of about 0.042 W/m K at room temperature and the conductivity of air at the same temperature is 0.026 W/m K. An increase in the packing density would increase the wool–air ratio and therefore increase the conductivity. A decrease in the packing density may decrease the effective conductivity towards 0.026 W/m K, but there is a lower limit owing to the commencement of convection when the air spaces become large.

For many engineering applications an insulation material with a degree of mechanical strength is required and materials such as furnace bricks, asbestos boards and cements, and foam are commonly used. It can be seen in Table 2.3 that some sacrifice of low conductivity is necessary if the material is to have rigidity. Expanded foams such as polystyrene have a fairly low conductivity and have the advantage in some applications that they may be expanded *in situ*. The conductivity increases with increase in temperature, as is to be expected because the conductivity of air (which is the main volumetric constituent) increases with increased temperature.

TABLE 2.3 The apparent thermal conductivity of some insulating materials in air

Material	*Thermal conductivity* (W/m K)			
	0° C	*100° C*	*300° C*	*500° C*
Asbestos felt	0.16	0.19	0.21	0.23
Mineral wool	0.047	0.060	0.097	0.14
Animal wool	0.038	0.058	—	—
Fireclay brick	0.55	0.6	0.7	0.8
Microporous silica	0.017	0.021	0.03	0.04
Polystyrene foam	0.03	—	—	—
Air (for comparison)	0.024	0.032	0.045	0.056

The reader may notice in Table 2.3 that one substance has a conductivity which is less than that of air, and in view of our discussion this may appear somewhat surprising. However, it is found that the conductivity in gases is

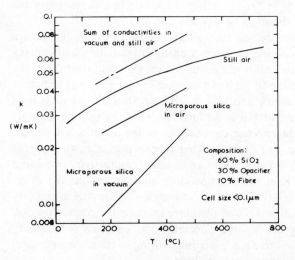

Figure 2.12 The conductivity of opacified microporous silica. (From Poole, 1967.)

dependent among other parameters on the molecular mean free path. Microporous silica has a pore size (about 0.1 µm) which is less than the mean free path in air, and it therefore effectively increases the resistance to heat flow in a similar way to the effect of crystal size in solids at low temperature mentioned in Section 2.3. Fig. 2.12 gives the conductivity variation with temperature and also shows that the conductivity in air is less than the sum of the conductivity in a vacuum (due to the matrix alone) and the conductivity of the air. In the absence of this physical shortening of the mean free path by the pore size and assuming no convection, the total conductivity would be approximately equal to the sum of the conductivities. The transmission of radiation through the material is reduced to a negligible quantity by the opacifier.

Heat Flow due to the Movement of Matter 3

3.1 The Thermal Skin

At its best a motor car is a very inefficient piece of machinery. Its engine may be capable of transferring about one-quarter of the energy in the fuel supplied to useful work under conditions of optimum loading, but under road conditions of acceleration and deceleration and part-load running this fraction falls to nearer one-tenth. The rest of the energy is nearly all lost as heat to the surroundings. Some of this heat is dissipated by the exhaust gases and by air cooling of the engine block but the greatest single outlet is through the water cooling system to the radiator. A motor car radiator may pass a quantity of energy to the surrounding air which is typically 3 or 4 times that produced as work by the engine, and it achieves this heat transfer with temperatures of less than the boiling point of water. Since, as we have seen in previous chapters, radiation at this temperature is low and conduction in air and water is low, the high heat transfer rate must be due to movement of the fluids; a process commonly termed convection.

It was mentioned earlier that convection is an indirect form of heat transfer and that radiation and conduction are the only basic heat flow processes. Convection typically involves heating a small mass of fluid, at or near a surface, by radiation and conduction and therefore increasing its temperature and internal energy. This mass then circulates into the bulk of the fluid due to general movement and transfers its excess internal energy to the bulk, once again by radiation and conduction. This complete process takes place in a very thin layer or skin adjacent to the surface and the characteristics of this skin form the basis of convection heat transfer theory. Convection is notable by its absence from textbooks and scientific writings of the nineteenth century. This is particularly surprising in view of its noble pedigree (Newton, 1701) and the fact that it is by far the most common and important form of heat transfer in engineering applications. Ironically there are now probably more research papers on convection than any other specialized scientific subject.

The thermal conductance of the skin or boundary layer is given by the heat transfer coefficient h as defined and used in Chapter 2:

$$q = \frac{Q}{A} = -h(T_2 - T_1) \tag{3.1}$$

The heat transfer coefficient varies with the fluid properties (such as the density, thermal conductivity, specific heat and viscosity), the velocity, the surface roughness and the thickness and type of boundary layer. This latter point may require some amplification for readers not conversant with the qualitative aspects of boundary layer formation. Consider the parallel flow of air over a flat plate as shown in Fig. 3.1. In this figure the ordinate scale is about 100 times

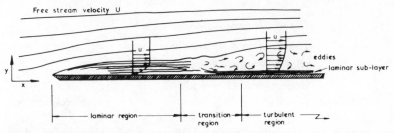

Figure 3.1 Boundary layer on a flat plate.

greater than the abscissa scale, the boundary layer generally being a fraction of a centimetre thick. Downstream from the front edge of the plate a laminar layer builds up in which the velocity, u, varies smoothly but not linearly from zero at the wall to the free stream velocity at the outer edge of the layer. The velocity gradient is proportional to the shear stress τ between successive fluid layers such that

$$\tau = \mu \frac{\mathrm{d}u}{\mathrm{d}y} \tag{3.2}$$

where the constant of proportionality μ is called the dynamic viscosity. At a certain distance along the plate the flow within the boundary layer becomes unstable, in the sense that any small perturbation within the laminar flow is not damped out but increases in amplitude. This causes the laminar flow to collapse gradually over a distance, termed the transition region, and results in a disordered flow with the swirls and eddies which characterize turbulence. At the surface another laminar layer emerges which is much thinner than the original laminar layer and is termed the sub-layer.

The major resistance to heat transfer is provided by this sub-layer and, as it is thin, it is to be expected that turbulent flow leads to high heat transfer coefficients. In practice this is most desirable because scientific and engineering applications generally involve turbulent flow. However, laminar flow with its smooth streamlines is more easily analysed theoretically and it does occur in natural convection, in the flow of condensate around condenser tubes and various other fields. The term 'natural convection' is reserved for convection caused by density gradients (generally as a result of temperature differences) within a fluid, as distinct from 'forced convection' where the bulk fluid motion is caused by external means such as a fan or pump. In this chapter on the fundamentals the reader will neither be bombarded with an analytical presentation of boundary layer theory nor submerged in a sea of empirical relationships, although both these aspects are important and are covered later. Instead, a description of a simple analogy with momentum transfer, and an introduction to dimensionless groupings is undertaken.

3.2 Eddies

> *Big whirls have little whirls,*
> *That feed on their velocity;*
> *And little whirls have lesser whirls,*
> *And so on to viscosity.*

L. F. Richardson

In more prosaic terms, random mixing in fluids involves a range of fluid movements extending from fairly large whirls and vortices down to microscopic molecular interactions. It is often convenient to divide a wide-ranging effect into two or more regions for the purpose of analysis. For this reason turbulent flow is considered as the combination of microscopic movements and general macroscopic movements which we shall term 'eddies'. In laminar flow there are, by definition, no eddies and the purely molecular movement leads to a shear stress and dynamic viscosity defined as in equation (3.2). For our purposes here, this equation is more conveniently expressed in terms of the kinematic viscosity v where $v = \mu/\rho$. Equation 3.2 therefore becomes:

$$\tau = \rho v \frac{du}{dy}$$

In turbulent flow the effect of the eddies may be treated as an additional viscosity, termed the eddy diffusivity (or turbulent diffusivity) ε such that:

$$\tau = \rho(v + \varepsilon) \frac{du}{dy} \tag{3.3}$$

The units of ε and v are m^2/s and in most turbulent flows ε is very much larger than v.

Turning our attention now to thermal aspects we can say that in laminar flow, where there is no fluid movement in the y direction, heat transfer from the wall must be by conduction (neglecting any radiation within the fluid). Fourier's law of conduction may be written as:

$$q = -\rho c \alpha \frac{dT}{dy}$$

where c is the specific heat of the fluid at constant pressure and α is the thermal diffusivity defined as $\alpha = k/\rho c$ (see Section 7.1). For turbulent flow an analogy may now be made with the viscous effects by assuming that the eddies effectively cause an additional and similar increase to the thermal diffusivity such that:

$$q = -\rho c(\alpha + \varepsilon) \frac{dT}{dy} \tag{3.4}$$

Similarly in most turbulent flows ε is very much larger than α.

Under conditions where $v = \alpha$ (or when they are approximately equal as both terms are generally much less than ε) division of equation (3.4) by equation (3.3) gives:

$$q \, du = -c\tau \, dT$$

Integration between the free stream velocity U and zero velocity at the surface, and over the temperature difference ΔT between the free stream and the surface yields:

$$q = -\frac{c \, \Delta T}{U} \tau_0 \tag{3.5}$$

This analogy therefore essentially gives a relationship between the heat flux and fluid shear stress, τ_0 at the wall, under conditions when $v = \alpha$ and is termed Reynolds analogy after the originator of the concept (Reynolds, 1874). The ratio v/α is termed the Prandtl number Pr after Ludwig Prandtl who introduced the concepts of boundary layer theory at the turn of the century:

$$\text{Pr} = \frac{v}{\alpha} = \frac{c\mu}{k}$$

The Prandtl number is a property of the fluid and Table 3.1 indicates its value for some common fluids. Reynolds analogy predicts reasonable results when the deviation of Pr from unity is moderate and it may therefore be used for many gas flow problems. A visualization of Reynolds analogy is given by Fig. 3.2 in which a hypothetical pocket of fluid at the wall is transferred into the free stream by eddies.

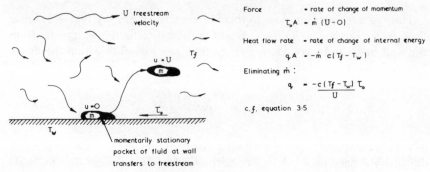

Figure 3.2   Visualization of Reynolds analogy.

TABLE 3.1   Some properties of common fluids

Fluid	T °C	ρ kg/m³	k W/m K	c_p kJ/kg K	μ 10^6 kg/m s	Pr $(c_p\mu/k)$
Air	20	1.21	0.0257	1.005	18.1	0.71
	100	0.94	0.032	1.01	21.8	0.69
Hydrogen	20	0.084	0.178	14.3	8.83	0.71
	100	0.063	0.216	14.4	10.4	0.69
Steam	100	0.60	0.025	1.9	12.2	0.93
Water	20	998	0.60	4.18	1002	7.0
	100	958	0.68	4.22	279	1.7
Freon-12 (liquid)	20	1330	0.073	0.97	270	3.6
Crude oil	20	∼900	∼0.1	∼2.0	∼50,000	∼1000
Mercury	20	13,500	8.0	0.14	1570	0.027

As an example of the practical application of the analogy, consider the flow of air at a temperature of 100 °C and velocity of 4 m/s along a tube of 25 mm diameter as shown in Fig. 3.3. The pressure drop per unit length $\Delta p/l$ is measured and found to be 11 N/m³. The approximate heat transfer coefficient may be estimated as follows.

From a force balance over length l

$$\Delta p \,\frac{\pi d^2}{4} = \tau_0 \pi dl$$

therefore

$$\tau_0 = \frac{\Delta p}{4}\frac{d}{l} \tag{3.6}$$

Also $q = -h\,\Delta T$ where ΔT represents the temperature difference between the fluid and the tube wall. Substitution in equation (3.5) then yields

$$h = \frac{c}{4}\frac{\Delta p}{U}\frac{d}{l}$$

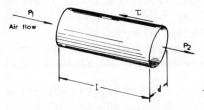

Figure 3.3 Air flow through a tube.

and numerical substitution (with the specific heat at constant pressure taken at 100 °C) gives

$$h = \frac{1.01 \times 11 \times 0.025}{4 \times 4} = 0.017 \text{ kW/m}^2 \text{ K}$$

This value of h is only approximate owing to the assumptions made in Reynolds analogy. One further assumption which was not mentioned earlier, for the sake of simplicity, is that the eddy diffusivity ε is common to both the viscous and thermal effects. This is not strictly the case and ε in equations (3.3) and (3.4) should more precisely be replaced by ε_m and ε_t respectively where the suffixes denote momentum and thermal forms of the diffusivity. By using this substitution in classical boundary layer theory, some progress has been made towards an entirely analytical treatment of turbulent flow heat transfer over wide ranges of fluid properties and flow conditions. In Chapter 8 the reader is introduced to the momentum, energy and Navier–Stokes equations of the boundary layer in preparation for more specialised or advanced work on convective heat transfer analysis.

3.3 Data Correlation

In the previous example of air flowing along a tube we were able to estimate the heat transfer coefficient from measurements of the pressure drop. More commonly these pressure drop measurements would not be available because the estimation of h is required at the design stage. It may then be necessary to search the literature and find a similar heat flow situation in which the pressure drop, or better still the heat transfer coefficient, has been measured. But perhaps measurements have never been made in which the fluid, the geometrical configuration and the temperatures are the same as those required. In fact, since the variable properties which affect the situation include c, μ, k, U, τ, d and ΔT, it is rather unlikely that in any particular practical case previous measurements under the same conditions will have been recorded. It is therefore advantageous to simplify the analysis by identifying groups of properties which can be combined and have a characteristic effect on the overall situation. One of these groups we have already introduced as the Prandtl number, $\text{Pr} = c\mu/k$. The groups are arranged to be dimensionless in order that the relationships between them are not a function of the measuring units employed.

47

The relationship obtained from Reynolds analogy (equation (3.5)) may be transformed into dimensionless parameters. In fluid mechanics the fluid shear stress, τ_0, is expressed in terms of the fluid velocity as

$$\tau_0 = \frac{f}{2} \rho U^2 \qquad (3.7)$$

where f is the friction factor (or friction coefficient) and is dimensionless. The friction factor is a measure of the resistance to flow and effectively relates the pressure drop to the velocity. In the case of circular tubes, this is shown by combining equations (3.6) and (3.7)

$$\Delta p = \frac{f}{2} \frac{4l}{d} \rho U^2$$

The friction factor defined by equation (3.7) is that conventionally used in convection heat transfer and is sometimes called the Fanning friction factor to distinguish it from the Darcy friction factor which is 4 times greater. Another important dimensionless group in fluid mechanics is the Reynolds number Re defined as

$$\text{Re} = \frac{\rho U d}{\mu} \qquad (3.8)$$

where the linear dimension d is conventionally the diameter in the case of a tube and the distance from the leading edge in the case of a flat plate. The Reynolds number relates the momentum and viscous forces and, as we shall see, provides an indication of the type of flow to be expected. Returning to Reynolds analogy (equation (3.5)), substituting from equation (3.7) for τ_0 and putting $-h\,\Delta T$ for q yields

$$h = \frac{f}{2} \rho c U$$

Rearrangement to include the Reynolds and Prandtl numbers yields

$$\frac{hd}{k} = \frac{f}{2} \left(\frac{\rho U d}{\mu} \right) \left(\frac{c\mu}{k} \right) \qquad (3.9)$$

The dimensionless group which results on the left-hand side is termed the Nusselt number Nu and relates the convective and conductive heat flow rates in the fluid. It should be noted that all the properties relate to the fluid and not the surface. Reynolds analogy may now be expressed as

$$\text{Nu} = \frac{f}{2} \text{Re Pr} \qquad (3.10)$$

The Reynolds number is used to distinguish between laminar and turbulent flow as shown in Fig. 3.4. (In the case of an external boundary such as that on a flat plate the frictional resistance is conventionally expressed in terms of an average drag coefficient C_d which replaces the friction factor f and is

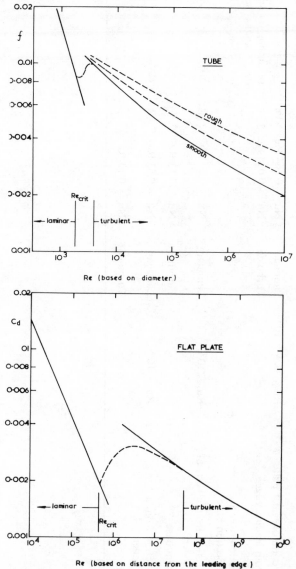

Figure 3.4 Friction coefficients.

effectively the frictional drag $\tau / \frac{1}{2} \rho U^2$ averaged over the plate length.) In both these geometrical arrangements there is a change in the relationship between the parameters which corresponds to the change from laminar to turbulent flow. In practice this change may occur anywhere within a range termed the transition region and its exact position depends upon the smoothness of the

TABLE 3.2 Some expressions for the Nusselt number (suitable for use under normal conditions and with Pr in the range $\frac{1}{2}$–100)

Flow configuration (and physical dimension used)	Approx. flow transition	Form of flow	Convection heat flow equation (based on mean heat transfer coefficient)
1. Forced convection through pipe d = pipe internal diameter	Re = 2300	Laminar	Nu = 4.36 (constant heat flux) Nu = 3.65 (isothermal wall)
		Turbulent	Nu = $0.023\,Re^{0.8}\,Pr^{0.4}$ (heating fluid) Nu = $0.023\,Re^{0.8}\,Pr^{0.3}$ (cooling fluid)
2. Forced convection over flat plate d = length from leading edge	Re = 0.5×10^6	Laminar	Nu = $0.664\,Re^{0.5}\,Pr^{0.33}$
		Turbulent	Nu = $0.036\,Re^{0.8}\,Pr^{0.33}$ (assuming no initial laminar section)
3. Forced convection over a cylinder—see Section 8.3.5			
4. Natural convection from a vertical surface or cylinder d = height from base	Ra = 10^9	Laminar	Nu = $0.59\,Ra^{0.25}$ (Ra > 10^4)
		Turbulent	Nu = $0.13\,Ra^{0.33}$
5. Natural convection from a horizontal cylinder d = diameter	Ra = 10^9	Laminar	Nu = $0.53\,Ra^{0.25}$ (Ra > 10^4)
		Turbulent	Nu = $0.13\,Ra^{0.33}$

surface and the flow. The critical Reynolds number Re_{crit} for tubes is generally taken as 2300 (with the linear dimension d equal to the diameter of the tube) and for flat plates as 500,000 (with d equal to the distance from the leading edge). The first step in many forced convection problems in which the flow is not obviously turbulent is to evaluate Re and hence determine the nature of flow.

The Nusselt number is generally the required dimensionless group in thermal convection problems as it includes the heat transfer coefficient; $Nu = hd/k$. In relationships involving the Nusselt number and other groups which contain a linear dimension d this dimension must be common between the groups. The Nusselt numbers for tubes and flat plates therefore involve the tube diameter and distance from the leading edge, respectively. Practical relationships for the Nusselt number are generally expressed in terms of Reynolds number and other dimensionless groups and are often non-linear. For example, the expression commonly used to estimate turbulent flow heat transfer in a tube (and replacing the approximate relationship given by equation (3.10)) is the Dittus–Boelter (1930) equation:

$$Nu = 0.023\ Re^{0.8}\ Pr^{0.4} \tag{3.11}$$

This equation represents the empirical data to within 25% for fluid being heated in a tube and is valid for fluids with a Prandtl number within the range 0.6 to 100. It may therefore be used for air and water flow for example, but not liquid metals or viscous oil. The exponent of the Prandtl number is reduced to 0.3 for flows in which the fluid is cooled. In the case of laminar flow in a tube under conditions of constant heat flux the heat transfer is independent of the flow as we shall see in Chapter 8. The Nusselt number is then a constant and it is found analytically that $Nu = 4.364$. These and similar expressions for other configurations are included in Tables 3.2 and 3.3.

TABLE 3.3   Some simplified expressions for the mean heat transfer coefficient for air (at approximately atmospheric pressure and temperature)

1. Forced convection through pipe, turbulent flow, velocity U	$(Nu = 0.02\ Re^{0.8})$	$h = 0.0037\ U^{0.8}/d^{0.2}$
2. Forced convection over smooth plate of length about 0.5 m	$U < 5$ m/s $U = 5$ to 30 m/s	$h = 0.005 + 0.004\ U$ $h = 0.007\ U^{0.8}$
3. Forced convection at right angles to cylinder, velocity U	$Re = 1000$ to $50{,}000$ $(Nu = 0.24\ Re^{0.6})$	$h = 0.005\ U^{0.6}/d^{0.4}$
4. Natural convection from horizontal cylinder, diameter d (m), ΔT in °C	$Ra = 10^4$ to 10^9 $Ra > 10^9$	$h = 0.0013\ (\Delta T/d)^{0.25}$ $h = 0.0012\ \Delta T^{0.33}$
5. Natural convection from vertical surface or cylinder, height l (m), ΔT in °C	$Ra = 10^4$ to 10^9 $Ra > 10^9$	$h = 0.0014\ (\Delta T/l)^{0.25}$ $h = 0.0013\ \Delta T^{0.33}$

The semi-empirical analysis of a complex subject by the formation of dimensionless groups is termed 'dimensional analysis'. It enables relationships between properties to be expressed in a manner which is not dependent on the system of units and is therefore particularly useful in scale modelling. The theory of similarity and the generalized formation of dimensionless groups is not covered here as it is well described in many text books on fluid flow and convection. There is only one remaining basic dimensionless group commonly used in convective heat transfer which we have not introduced and that is the Grashof number. This group arises as an important parameter in natural or free convection and is defined as

$$\mathrm{Gr} = \frac{g\beta \, \Delta T d^3}{v^2} \tag{3.12}$$

where g is the force per unit mass due to gravity (or other body force), β is the coefficient of volumetric expansion, ΔT is the temperature difference between the surface and the bulk fluid, d is a linear dimension and v is the kinematic viscosity. The Grashof number is the ratio of the buoyancy force to the viscous force in natural convection and plays a similar role to Reynolds number in forced convection.

Fluid flow due to natural convection may be laminar or turbulent in a similar way to flow in forced convection. This is well illustrated by the rise of cigarette smoke in still air, where the flow is initially laminar but at a height of about 200 mm becomes turbulent. Heat transfer by natural convection is often empirically correlated by a relationship of the form

$$\mathrm{Nu} = C(\mathrm{Gr} \times \mathrm{Pr})^m$$

where C and m are constants. The product $\mathrm{Gr} \times \mathrm{Pr}$ leads to a derived dimensionless group termed the Rayleigh number, Ra, and some expressions for Nu in terms of Ra are included in Table 3.2. Other derived groups include the Stanton number, St, and the Peclet number, Pe, which is used in forced convection involving liquid metal as the fluid. A summary of the dimensionless groups is given in Table 3.4. (It is worthy of note that Reynolds analogy effectively equates the Stanton number to $f/2$.)

3.4 High Heat Transfer Coefficients

Considerably higher heat transfer coefficients than those associated with the convective situations we have been discussing are obtained in processes such as boiling, fluidization and condensation. In the case of boiling, under conditions when a steady stream of bubbles are produced, the heat exchange due to the violent turbulence at the surface and vaporization of the liquid leads to a heat transfer coefficient which may be up to a few orders of magnitude greater than that found in forced convection. Fluidization involves passing a

TABLE 3.4 Dimensionless groups used in convective heat transfer

Reynolds number	$Re = \dfrac{\rho U d}{\mu}$
Nusselt number	$Nu = \dfrac{hd}{k}$
Prandtl number	$Pr = \dfrac{c_p \mu}{k}$
Grashof number	$Gr = \dfrac{g\beta \,\Delta T d^3}{\nu^2}$
(Stanton number)	$St = \dfrac{h}{\rho c_p U}$
	$\quad = \dfrac{Nu}{Re \times Pr}$
(Peclet number)	$Pe = \dfrac{\rho c_p U d}{k}$
	$\quad = Re \times Pr$
(Rayleigh number)	$Ra = \dfrac{g\beta \,\Delta T d^3 c_p \rho^2}{\mu k}$
	$\quad = Gr \times Pr$
Friction factor or drag coefficient	$f \text{ or } C_d = \dfrac{\tau}{\frac{1}{2}\rho U^2}$

fluid (liquid or gas) through solid particles such that their weight is supported by the fluid flow. Under this condition the suspended particles act as a liquid and find their own level in a container. The heat transfer coefficient between the fluid and the container walls is large as the violent agitation of the particles effectively reduces the boundary layer thickness at the wall. Condensation coefficients are high owing to vapour condensing on to a thin layer of condensate and transferring the latent heat to the surface. Each of these

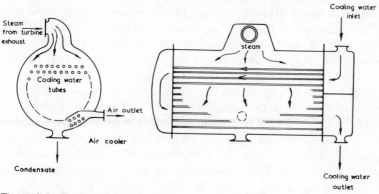

Figure 3.5 Typical two-pass condenser arrangement.

processes will be examined in Chapter 9 but for the moment let us qualitatively consider the condensation process as this will additionally raise one or two points in readiness for Chapter 5.

A typical arrangement of a power station shell and tube condenser is shown in Fig. 3.5. Steam from the turbine exhaust enters at the top and condenses

TABLE 3.5   The magnitude of convective heat transfer coefficients

	$h(\text{kW/m}^2\,\text{K})$
Forced convection	
surface to gas	0.01 –1
surface to liquid	0.1 –10
surface to liquid metal	1 –50
Free convection	
surface to gas	0.0005–1
surface to liquid	0.1 –5
surface to air with $\Delta T < 100\,°\text{C}$	0.001 –0.01
Nucleate boiling	0.5 –50
Filmwise condensation	0.5 –50

on the tubes which are cooled by cooling water from a river, the sea, or from cooling towers. The condensing steam forms a layer of liquid around the tube and a dynamic equilibrium exists between this continually renewed layer and the condensate dripping off the bottom of the tube. This thin layer of liquid forms the main resistance to heat transfer and the local heat transfer coefficient is higher at the top of the tube than at the bottom where the layer is thickest. An analysis of this condensate layer, originally formulated by Nusselt in 1916, leads to a successful relationship for the condensation heat transfer coefficient and is presented in Chapter 9.

In practice the condensate does not always form a smooth film around the tube. Ripples may occur and at high flow rates turbulence may commence. Grease and dirt may prevent the surface becoming wetted so that the condensate appears as drops (rather like rain on a window pane) and this type of condensation is termed 'dropwise' to differentiate it from 'filmwise' condensation. Dropwise condensation is advantageous as the heat transfer coefficients can be up to an order of magnitude higher than those achieved in filmwise condensation. This occurs because most of the vapour condenses between the drops where the liquid layer thickness may be microscopic. Unfortunately, agents added to prevent non-wetting are eventually washed away and condensation reverts to the filmwise form. A considerable amount of research has been conducted on the problem of maintaining dropwise condensation but no entirely successful solution has yet been found. It must also be borne in mind that there is no great advantage to increasing the condensation heat transfer coefficient unless the cooling water side coefficient is also increased, because the overall coefficient is dependent on both of these coefficients.

An analysis of the condenser as a whole must involve the definition of mean temperatures which can be substituted in the heat transfer equation. The cooling water flowing through a tube will, at any particular tube section, be warmer at the tube wall than at the centre and a mean flow temperature must therefore be established. Furthermore the temperature difference between the cooling water and the condensing steam will decrease along the tube in the direction of flow and again an average value must be established for the purpose of analysis. This analysis is undertaken in Chapter 5 under the heading of 'heat exchangers' which is a general term taken to include boilers, condensers, cooling towers, car radiators, furnaces and any other system which has the function of transferring energy from one fluid to another. The condensing heat transfer coefficient must also be expressed as an average value for practical purposes. In the case of a condenser tube this means that the local heat transfer coefficient must be integrated around the circumference of the tube to give a mean value. Table 3.5 is included to give some appreciation of the magnitude of the mean heat transfer coefficients for various arrangements in SI units.

Heat Flow from Human Beings

4

The thermal comfort of human beings is one of the most important areas in which the fundamental modes of heat transfer discussed in the previous chapters can be directly applied. Until recent times the open fire provided the only means of keeping warm in cool climates and the deeply rooted intrigue of fire has not been entirely obliterated by modern society. Electrical heater manufacturers, for example, still feel the need to supplement the heat source by artificial glowing coal and flickering flames. But, in general, the days of fire as a centre of warmth and social life are gone and the simple process of keeping warm has now furcated into the more sophisticated areas of central heating, building insulation, air conditioning, clothing materials, district heating and double glazing. Well over half the total fuel consumed in Britain is used for heating human beings in either domestic or commercial and industrial premises. It is therefore important in terms of general energy conservation to reduce the heating load. In this chapter the factors involved in a body heat balance and the concepts of thermal comfort are studied. Finally, some practical conclusions are drawn regarding topics such as clothing, double glazing and central heating. The quantitative data in this chapter have largely been gathered from ASHRAE (1967) and the excellent text on thermal comfort by Fanger (1970).

4.1 Thermal Contact

It is common experience that material such as wool feels warm to touch and metal feels cold. Since this is the case when both the wool and the metal are at room temperature it follows that thermal feel must be dependent on properties other than temperature alone. In this section we shall attempt to analyse this situation and determine these properties.

The internal temperature of the human body is about 37 °C and the mean skin temperature of the whole body under conditions of low activity is about 34 °C. The skin temperature of the hands and feet tends to be rather below this value but for this discussion it will be assumed that all the skin is nominally at 34 °C. There is therefore a heat loss to the surroundings which is governed by the overall heat transfer coefficient between the skin surface and the surrounding air. This heat transfer coefficient is very sensitive to the clothes covering the skin and to air movement caused by activity of the body or draughts. In order to avoid these complications for the moment, we shall consider a finger situated in still air as shown in Fig. 4.1a. If it is assumed that the finger is in thermal comfort in surroundings at 20 °C, and has a skin temperature of 34 °C, the heat transfer coefficient due to natural convection in still air may be calculated from part 5 of Table 3.3 (with $l = 0.06$ m) as about 0.0055 kW/m² K. The heat flux due to convection from the skin may be calculated from the product of h and the temperature difference as 0.077 kW/m² and together with a radiation flux would yield the total heat flow from the finger per unit area.

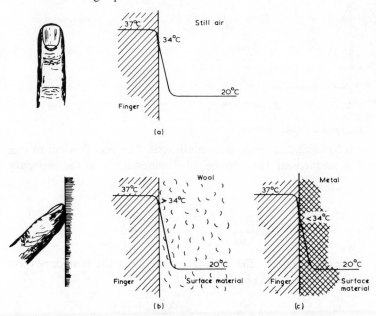

Figure 4.1 Thermal feel.

When the finger touches a surface at room temperature, as shown in Figs. 4.1b and 4.1c, the heat flow from the skin in contact with the surface may increase or decrease depending on the properties of the surface material. The problem involves transient conduction as the temperature of the surface material gradually increases to an equilibrium value. If the material is wool insulation, the initial heat flux will be reduced below the heat flux in still air, and the skin temperature will rise slightly above 34 °C leading to a sensation of warmth. This is experienced when a glove is put on. If, on the other hand, the surface material is a metal the initial heat flux will be increased, the skin temperature will fall slightly and the metal will feel cold. The schoolboy definition of temperature as 'the degree of hotness' is therefore rather ambiguous if 'hotness' applies to the sense of feel.

An approximate relationship between the temperature differences and the relevant properties may be found as follows. If two materials with different temperatures are put in contact, the temperature distribution after a very short time will be as shown in Fig. 4.2 with a thin layer at the surface of each

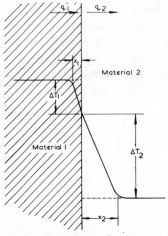

Figure 4.2 Thermal contact analysis.

material which has adjusted to the new conditions. The heat flux out of one material will equal the heat flux into the other material and at the boundary therefore:

$$q_1 = q_2$$

That is,

$$-k_1\left(\frac{\mathrm{d}T_1}{\mathrm{d}x_1}\right)_{x=0} = -k_2\left(\frac{\mathrm{d}T_2}{\mathrm{d}x_2}\right)_{x=0}$$

Using the approximation that the thermal gradients at the boundaries are given by $\Delta T_1/x_1$ and $\Delta T_2/x_2$ as shown in Fig. 4.2:

$$-k_1\frac{\Delta T_1}{x_1} = -k_2\frac{\Delta T_2}{x_2}$$

In addition, the accumulative heat energy stored in material 2 is equal to the total heat lost from material 1. If the approximation is made that the temperature distribution in the layer of thickness x is linear, the mean temperature difference each side is $\frac{1}{2}\Delta T_1$ and $\frac{1}{2}\Delta T_2$, and the equality becomes

58

$$x_1\rho_1c_1\left(\frac{\Delta T_1}{2}\right) = x_2\rho_2c_2\left(\frac{\Delta T_2}{2}\right)$$

where ρ and c denote density and specific heat. Elimination of x_1 or x_2 between these equations yields

$$\rho_1c_1k_1(\Delta T_1)^2 = \rho_2c_2k_2(\Delta T_2)^2$$

from which

$$\frac{\Delta T_1}{\Delta T_2} = \frac{\sqrt{(\rho_2c_2k_2)}}{\sqrt{(\rho_1c_1k_1)}} = \frac{b_2}{b_1} \tag{4.1}$$

where b is the contact coefficient defined as $\sqrt{(\rho ck)}$. The contact coefficients for various materials are given in Table 4.1 and are arranged in order of the coldness felt on touching them. Account of the contact coefficient is sometimes

TABLE 4.1   The contact coefficient (at $20\,^{\circ}$C)

Material	Contact coefficient $b(kJ/m^2\ K\ s^{1/2})$
Copper	36
Mild steel	15
Concrete	1.6
(Water)	1.6
Wood, oak	0.35
Asbestos board	0.28
Wool	0.039
Polyurethane, expanded	0.032

taken when selecting materials for chairs, bathroom floors, door handles and other items. For example, it may not be possible to hold a metal-surfaced car steering wheel in comfort on a hot day when the sun has been shining into the car. A covering of plastic or wood on the wheel will alleviate this problem by reducing the contact coefficient.

4.2   Thermal Balance

The thermoregulatory system of the human body maintains a constant temperature and there is therefore no net heat storage. The chemical energy supplied to the body may be equated to the heat and work output as shown in Fig. 4.3. The metabolic rate M is the total energy released by the oxidation processes in the human body per unit time. The total heat flow rate Q dissipated by the body may be considered as the sum of the heat flow rate $\dot{Q}_m$, due to *mass exchange* effects such as sweating and respiration, and the heat flow rate Q_h, due to *heat transfer* from the skin through any clothing to

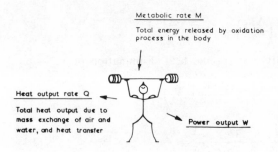

Figure 4.3 The human engine.

the surroundings. A balance on the human engine then yields

$$M - W - Q_m - Q_h = 0 \tag{4.2}$$

An efficiency μ of human activity is defined as W/M where W is the total physical work rate or power. Expression of this equation in terms of unit body surface area and rearrangement then gives;

$$\frac{M}{A}(1 - \mu) - q_m - q_h = 0 \tag{4.3}$$

where q_m is the heat flow rate per unit area of skin due to sweating, evaporation and respiratory effects

q_h is the heat flow rate per unit area of skin due to heat transfer from the skin and

A is the total skin area (throughout this chapter).

We shall now proceed to examine the terms of equation (4.3) in greater detail.

The magnitude of the metabolic rate is a function of the body activity and Table 4.2 gives values of M/A and μ for some typical activities. The mean nude body area of human beings is 1.77 m^2 (Fanger, 1970) and this value has been used in calculations throughout the chapter. A person walking up a gradient of 1:4 at 3.2 km/h (2 mph) therefore generates a power of 1.77 × 390 × 0.21 = 145 watts or about $\frac{1}{5}$ of a horse-power and the heat generated is 545 watts. It is apparent that in normal situations, where people are seated or occupied with a light activity, the heat generated is about 120 watts per person. The heating load required for a room may therefore be considerably reduced when a large number are present.

The heat loss q_m per unit area due to mass exchange effects may be sub-divided into four parts:

$$q_m = q_d + q_e + q_a + q_{wa} \tag{4.4}$$

where q_d is the heat flux due to water *diffusion* through the skin to the surroundings

TABLE 4.2   The metabolic rate and work efficiency for various activities

Activity	$M/A(\mathrm{W/m^2})$	μ
Sleeping	40	0
Sitting quietly	60	0
Driving car	60	0
Typing	70	0
Sitting—heavy limb work	130	0–0.2
Walking on level at 4.8 km/h	150	0
Walking up 1:20 gradient at 4.8 km/h	230	0.11
Walking up 1:4 gradient at 3.2 km/h	390	0.21
Pushing and handling wheelbarrow	150	0.2
House cleaning	120–200	0–0.1
Handling 50 kg bags	230	0.2
Sawing wood	250	0.1–0.2
Playing tennis	270	0–0.1
Playing squash	420	0–0.1

q_e is the heat flux due to *evaporation* of water or sweat at the surface of the skin

q_a is the heat flux (based on total skin area) due to heating of *air* above the surrounding temperature by the respiratory system

q_{wa} is the heat flux (based on total skin area) due to evaporation of *water in the air* exchanged by the respiratory system

These terms are all connected with the exchange of air or water or the evaporation of water, and each may be expressed as the product of a mass flow rate and enthalpy change. Typical values calculated for various conditions are presented in Table 4.4.

The last term of equation (4.3), q_h, refers to the heat flux through the clothing by conduction and then to the surroundings from the outer surface

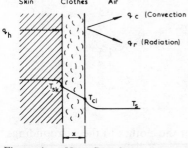

Figure 4.4   Heat flow from the skin.

of the clothing by convection and radiation (as shown in Fig. 4.4). In terms of conduction through the clothing it may be expressed as

$$q_h = -\frac{A_c}{A}\frac{k}{x}(T_{cl} - T_{sk})$$

61

where A_c is the area of clothing, (by convention the outer surface area of the clothing)

k is the apparent thermal conductivity of the clothing

x is the mean thickness of the clothing

T_{cl} is the temperature at the outer surface of the clothing

T_{sk} is the mean skin temperature.

The ratio A_{cl}/A is termed the area factor for the clothes, f_{cl}, and effectively corrects the equation for the fact that q_h is based on the area, A, of the body rather than the clothes. The term $Ax/A_c k$ is the total thermal resistance of the clothing and has units of $m^2\ K/kW$. It is usually expressed either as a 'clo'-index, I_{cl}, or a 'tog' where conversions from clo-units and tog units are:

$$1\ \text{clo-unit} = 155\ m^2\ K/kW$$

$$1\ \text{tog} \quad = 100\ m^2\ K/kW$$

therefore

$$1\ \text{tog} \quad = 0.645\ \text{clo-units}$$

I_{cl} is defined such that for a typical summer, lightweight business suit it has a value of unity. The heat transfer through the clothes is therefore given by

$$q_h = \frac{-(T_{cl} - T_{sk})}{155\ I_{cl}} \tag{4.5}$$

Some typical values of I_{cl} (and f_{cl}) are given in Table 4.3. The heat flow by

TABLE 4.3   The thermal resistance of clothing

Clothing	I_{cl} (clo)	f_{cl}
Nude	0	1.0
Shorts, casual shirt and sandals	0.3	1.05
Long trousers, cotton shirt and shoes	0.5	1.1
Summer lightweight business suit or casual wear and woollen sweater	1.0	1.15
Lightweight business suit and cotton coat	1.5	1.15
Heavy business suit with waistcoat, heavy underwear and woollen socks	1.5	1.2
As above with heavy woollen overcoat	3	1.3

convection and radiation from the surface of the clothes to the surroundings at temperature T_s is given by

$$q_h = q_c + q_r$$
$$= -(h_c + h_r)f_{cl}(T_s - T_{cl}) \tag{4.6}$$

The heat transfer coefficients h_c and h_r have been discussed in previous chapters, h_c in particular is sensitive to the air velocity. This is very evident from the strong cooling effect experienced in a wind or draught. This equation may be combined with equation (4.5) to yield the heat flux in terms of the overall temperature drop:

$$q_h = \frac{-f_{cl}(T_s - T_{sk})}{155\, I_{cl} f_{cl} + 1/(h_c + h_r)}\ \mathrm{kW/m^2} \qquad (4.7)$$

Some typical values of the parameters involved in a heat balance on the body for various activities are given in Table 4.4. The skin temperature and other experimental parameters were estimated from data given in Fanger (1970) and ASHRAE (1967); the heat flux values were calculated by assuming surroundings at 20 °C and a relative humidity of 50%. The convection heat transfer coefficient was estimated using reasonable values of the relative local air velocity for the activity concerned and the radiation coefficient was calculated by using equation (1.7) and assuming the emissivity of the clothing was unity. For a true heat balance the values of the last line should correspond to the values of the first line (equation (4.3)); discrepancies are due to the approximations involved and the rough estimation of the proportion of energy utilized as work.

TABLE 4.4 Typical body heat balance values under various conditions. (Heat flow rates are expressed in W/m² of skin surface and should therefore be multiplied by the mean body area of 1.77 m² to obtain typical heat flow rates for the total body)

Activity		Sleeping	Seated quietly	Walking on level	Handling 50 kg bags	Playing squash
Clothing		Blankets	Heavy suit	Light suit	Overall	Shorts only
$\dfrac{M}{A}(1-\mu)$	W/m²	40	60	150	180	380
T_{sk}	°C	34.6	34.0	31.6	30.7	~26
q_d	W/m²	20	19	18	17	14
q_e	W/m²	0	0	40	56	135
q_a	W/m²	1	1	3	5	8
q_{wa}	W/m²	3	4	12	20	40
q_m	W/m²	24	24	73	98	197
I_{cl}	clo	~4	1.5	1.0	1.0	~0
f_{cl}		~1.4	1.15	1.15	1.2	1.0
h_c	W/m² K	5	5	12	12	15
h_r	W/m² K	6	6	6	6	6
q_h (from equation (4.7))	W/m²	21	45	57	53	125
$q_m + q_h$	W/m²	45	69	130	151	322

4.3 Thermal Comfort

Thermal comfort involves creating an environment such that the maximum number of people in that environment would recommend no adjustment to the temperature or relative humidity. Many people spend over 90% of their lives in an artificial climate and it is therefore advantageous for both medical and psychological reasons to aim at optimum conditions.

If it is assumed that it is impossible to regulate externally the metabolic rate or the respiratory properties of human beings, the only major areas which offer scope for adjustment of the body's thermal system are the clothes and the convective and radiative heat transfer at the surface of the clothes. Basically the variable parameters must be optimized such that the mean skin temperature resulting from the overall heat balance is comfortable. This mean skin temperature is found to vary from about 35 °C for seated people to about 30 °C under conditions of heavy physical work. We shall now consider the influence of clothing and surrounding air velocity on the body heat balance.

The heat flow to the surroundings q_h is very sensitive to the clo-index as can be seen from Table 4.4. If, for example, a man were to change from wearing a heavy suit to a light suit, Table 4.3 indicates that the clo-index would fall from 1.5 to unity and, under conditions when he is seated quietly, substitution of relevant terms in equation (4.7) indicates that q_h would increase by 31%. Alternatively, if the metabolic rate remained at the same value, the temperature of the surroundings in which he was comfortable would increase from 20 °C to about 25 °C. In fact the clo-index is by far the most important parameter in determining the comfort temperature for any particular condition. The next most important variable is the mean air velocity. If the same man was seated under similar conditions to those in Table 4.4 but in a draught of air flowing at 1 m/s and at 20 °C, the value of h_c would increase to 0.012 kW/m² K and the comfort temperature would increase by about 2 °C.

Finally, as a rough guide, Table 4.5 has been included to show the temperature required for comfort under various conditions of clo-index and mean air velocity. Other less sensitive parameters are assumed to be normal with, for example, the relative humidity at 50% and the mean radiation temperature about the same as the surrounding temperature. Under normal conditions in 'still' air the mean relative air velocity due to general circulation will be of the order of 0.1 m/s and this value is therefore selected as the lower limit in the table. With the greater activity involved in manual work this value has been increased to 0.2 m/s.

4.4 Keeping Warm Economically

How can I reduce the cost of heating my home? In times of rising energy

TABLE 4.5   The comfort temperature

Activity	M/A (Table 4.2) (W/m²)	Clothing I_c (Table 4.3) (clo)	Relative air velocity (m/s)	Comfort temperature (°C)	(°F)
Sedentary	60	0	0.1	29	84
(seated with little		0.5	0.1	26	79
activity)		0.5	1.0	28	82
		1.0	0.1	23	73
		1.0	1.0	25	78
		1.5	0.1	21	70
		1.5	1.0	23	73
		3.0	0.1	~13	~55
Low	120	0	0.1	24	75
(standing and		0.5	0.1	20	68
walking about)		0.5	1.0	23	73
		1.0	0.1	15	59
		1.0	1.0	18	64
		1.5	0.1	11	52
		1.5	1.0	14	57
Medium	180	0	0.2	21	70
(general sustained		0.5	0.2	14	57
manual work)		0.5	1.0	19	66
		1.0	0.2	8	46
		1.0	1.0	12	54
		1.5	0.2	2	36
		1.5	1.0	6	43

costs this question is not only important to the householder but to society generally, as the major part of the total energy consumption is used for domestic heating. The question is also unanswerable in detail unless it is qualified by stating the type of house involved (or for that matter the office block or factory), the capital outlay available, the geographical position and many other parameters. Nevertheless, it will be found useful to examine the philosophy behind the question and to ascertain some helpful generalities regarding domestic heating.

Presumably the main concern of our questioner is with the heating of the occupants of his home and not the house itself, as there is no evidence that bricks and mortar have any particular thermal preference. A valid answer therefore would be for all the occupants to wear overcoats. Reference to Table 4.5 shows that use of a woollen overcoat (with I_{cl} of approximately 3 clo) would allow thermal comfort in surroundings at about 13 °C and would therefore considerably reduce the amount of heating required. In bed, under woollen blankets, thermal comfort may be maintained with external conditions below freezing. Although the extreme of wearing an overcoat is unacceptable, it is a fact that a considerable saving in heating load may be

65

achieved by a fairly modest increase in the clothing index. For example a man wearing an open-necked shirt and trousers or a woman in a light summer dress ($I_{cl} = 0.5$) will require an external temperature of about 26 °C when seated, while the addition of a jacket, woollen pullover or cardigan ($I_{cl} = 1.5$) will reduce the required temperature for comfort to 21 °C. This temperature drop may lead to a decrease in heating load that is more than directly proportional to the temperature change. This is because a large proportion of the heating load of the house is dissipated to the surroundings by natural convection, and for natural convection $Q \propto \Delta T^{5/4}$. The approximate saving in heat loss when the house temperature is lowered from condition 1 at 26 °C to condition 2 at 21 °C and when the atmospheric temperature is say 5 °C is therefore:

$$\frac{Q_1 - Q_2}{Q_1} = \frac{\Delta T_1^{5/4} - \Delta T_2^{5/4}}{\Delta T_1^{5/4}} = \frac{(26 - 5)^{5/4} - (21 - 5)^{5/4}}{(26 - 5)^{5/4}} = 29\%$$

Passing on from body insulation to the insulation of buildings, there are one or two general points that may be made. Heat is dissipated from buildings through the structure, the windows and by air exchange. In a draughty home the latter may well be the major factor. A typical six-roomed detached house may have a total volume V of 350 m³ and an air exchange rate equivalent to one complete change of air every half hour. When the room and atmospheric temperatures are 20 °C and 5 °C, the heat loss, Q_1, due to dry air exchange would be:

$$Q_1 = \rho \frac{V}{\Delta t} c_p \Delta T$$

$$= 1.22 \times \frac{350}{0.5 \times 3600} 1.0 \times 15$$

$$= 3.56 \text{ kW}$$

A small air exchange rate of around $\frac{1}{2}$–1 change/hour is required to keep the air fresh (Diamont, 1964) and there is therefore an unavoidable loss. The limiting of draughts through windows, under doors and down unused chimneys to this level is easy and inexpensive and could in the case above save 2 kW of power. Furthermore draughts lead to an uneven distribution of cooling load, as anybody who has sat in a room with a large gap under the door and consequently suffered cold feet will confirm.

The prevention of heat flow through the structure by filling in cavity walls, for example, is more difficult and expensive than draught exclusion (unless of course it is specified when the house is built). An evaluation of the capital cost and the annual saving would have to be undertaken to ensure the advisability of this work. Insulating the floor of the attic can be the most cost effective treatment of this type. Double glazing has become very popular in recent years partly owing to a vast publicity campaign which often includes misleading advertisements. Here again, a cost effectiveness study is required for detailed results but the conclusions resulting from the following case will be valid for many domestic properties.

A typical, modern, six-roomed detached house with no double glazing allows only 13–17% of the total heat dissipated (including air exchange) to flow through the windows. The fitting of double glazing in the situation shown in Fig. 4.5 will reduce the overall heat transfer coefficient by 57% and (if the

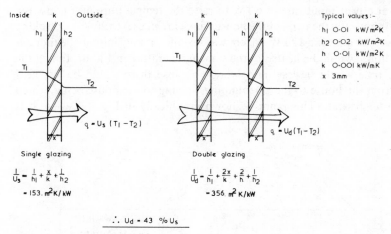

Figure 4.5 Double glazing.

heat flow through the windows is taken as 15% of the total heat dissipation) the saving in fuel cost will be 8.5%. The cost of double glazing is often of the order of a few hundred pounds sterling and a cost evaluation shows that this amount would provide the extra fuel required to cover the proposed gain from double glazing for 20 years or more. This reasoning is, of course, based purely on overall thermal grounds and consideration of noise abatement, more even distribution of temperature within the house and other factors may outweigh the economic considerations. Nevertheless, advertisements which claim a total fuel saving of one-third due to the fitting of double glazing are obviously exaggerated and misleading.

Central heating systems offer scope for economies by the careful regulation of temperature and distribution of heating load, in addition to the advantages mentioned earlier of lowering the overall house temperature and wearing a little extra clothing. Under conditions where there is some activity or movement such as walking about, the temperature required for comfort is lower than when sitting. Thus the kitchen, hallway and possibly the dining room require less heating than the sitting room. Bedrooms require less heating because the clo-index of bedding is usually high.

It is sometimes suggested that there is no saving in fuel consumption by intermittently heating and cooling a centrally heated house rather than leaving the heating on continuously. This is untrue, as any reduction in house temperature over some period must lead to a saving of fuel. Nevertheless it is true that there is not a saving which is directly proportional to the time the heating is off, owing to the thermal storage effect of the house structure. For this reason, in situations where it is not necessary to heat the house continuously (for example when the heating is turned off each night), there can be advantage in having a high capacity heating system which will heat the house quickly when required. As an illustration of this consider a system which requires 2 hours at a heat input rate of 6 kW to raise the temperature of a house (or more economically the parts of the house which are in use) to 20 °C at a waking time of say 7.0 a.m. If a 12 kW system was installed it may initially be thought that the house would be heated to the same temperature in 1 hour. In fact the required time would be less than 1 hour because there would be less heat passed from the house to the surroundings, owing to the house being heated in a shorter period. (This point is shown graphically in Fig. 4.6 where it is

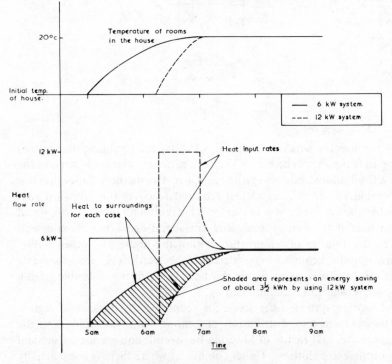

Figure 4.6 Central heating start-up characteristics.

assumed that the house temperature of 20 °C, once obtained, is maintained by the temperature setting of the central heating controller. The heat required will therefore reduce asymptotically towards the steady-state heat requirement of the house for the existing external conditions.)

Any attempt to place the fuels suitable for central heating systems, such as gas, coal, oil, direct electricity and 'off-peak' electricity, in an order of merit must take account of the fuel cost, availability, capital cost of the system and aesthetic considerations as well as thermal aspects. However, there are one or two general points concerning heat transfer which should be borne in mind. All fuels except electricity involve combustion and hence an unavoidable loss due to the exhaust of combustion gases. Consideration of fuels must therefore take into account the burning or boiler efficiency which may vary from about 60% to 85%. Electricity involves no loss in its conversion to useful heat but is more expensive per unit energy owing to the high proportion of heat loss in the thermodynamic generation cycle. It is obviously advantageous in terms of energy conservation to burn fuel directly for heating rather than use it to inefficiently generate electricity and use that for heating. As fuel becomes more scarce and therefore more expensive the relative costs of electricity and the other fuels must reflect this situation. In the long term, a case could be argued for the use of fuels other than electricity for space heating.

High temperature radiant heating from direct gas and coal fires or from radiant electric fires has the advantage that heat is transferred more quickly and directly to the occupants of a room than is the case with a mainly convective system. A person entering a cold room and turning on a radiant gas fire will benefit from an almost immediate heat transfer to his body, while a person switching on a hot-water convective central heating system may have to remain cold until the water in the system and the air in the room are heated. With intermittent heating requirements, high temperature radiant fires offer a considerable advantage. Since it is the occupants and not the house which generally require heating, the positioning of heating appliances is important. It is obviously extremely wasteful to have a single heating unit one end of a room (especially next to a window) and habitually sit at the other end. A further advantage of the high temperature radiant heater is its general positioning in a room such that the occupants are encouraged to sit around it in the radiant glow and tolerate lower temperatures in the extremes of the room.

Heat Flow in Exchangers

5

5.1 Enthalpy Exchange

Instruction in any scientific topic is limited to some extent by the language in which it is expressed. But surely, some may say, mathematics is absolutely precise and logical and therefore the perfect code for expressing scientific ideas. It is probably true that all valid scientific concepts can somehow be expressed mathematically but it does not follow that this is always the best method of demonstrating these concepts. The rules of bridge or chess, for example, can be easily explained and absorbed by word of mouth while a mathematical exposition would be complicated and tedious. In the fields of thermodynamics and heat transfer it is particularly important that both the intuitive feel and understanding of the subject, and the mathematical concepts and procedures, are advanced together. In fact one of the greatest analytical thermodynamicists, J. W. Gibbs, is credited with the statement that 'Mathematics is only a language'. On the other hand, the very fact that the meaning of words is not clearly defined has led to misunderstandings and in this field the word 'heat' is a good example.

Heat is the form of energy which flows due to a temperature difference and the important point for our present discussion is that it is only in evidence when it is flowing. It can never be latent or stored in any way and terms such as latent heat and heat storage are strictly misnomers. Heat flowing to a solid metal can lead to the metal melting because the increase in energy contained within the structure is sufficient to cause breakdown into the more disordered state found in a liquid; i.e., it is the increase in internal energy which has caused the melting. Heat supplied to an off-peak storage heater increases the temperature and therefore the internal energy of the storage material but it is not strictly stored as heat, although the loose terminology is generally tolerated owing to the common usage of these terms. A similar vagueness in the meaning of 'heat' is also very apparent in the field of heat transfer and terms such as heat loss, heat source and also heat exchange could be added to the list of misnomers.

A heat exchanger is a process component which has the primary object of transferring the energy stored in one fluid to another fluid. An analysis of the steady-flow energy equation, as found in most elementary thermodynamics text books shows that the property which indicates the total energy stored in a flowing fluid is the enthalpy. The rate of change of enthalpy in a fluid may be expressed as the product of the mass flow rate $\dot{m}$, the specific heat (at constant pressure) c and the temperature difference and, under normal conditions when no work is performed, is equal to the rate of heat transfer Q to or from the fluid. For a heat exchanger transferring energy from a hot fluid, suffix h, to a cool fluid, suffix c:

$$Q = -\dot{m}_h c_h \, \Delta T_h = \dot{m}_c c_c \, \Delta T_c \qquad (5.1)$$

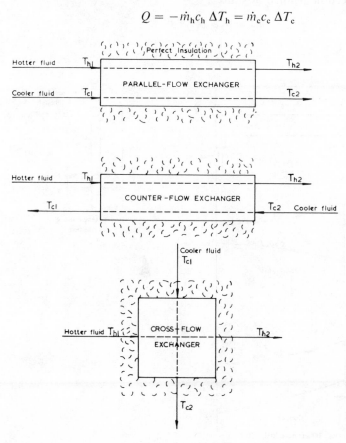

Figure 5.1 Some exchanger arrangements.

where the temperature difference ΔT is the outlet less the inlet temperature. It is assumed that the specific heat is constant with respect to temperature and in practice a mean value is used. It should be noted that the suffix notation used for heat exchangers throughout this text is as defined in Fig. 5.1 with h and c referring to the hotter and cooler fluids and 1 and 2 referring to the left-hand and right-hand (or top and bottom) sides respectively. The terms 'parallel-flow' and 'counter-flow' are generally used for describing the flow directions encountered in arrangements such as enclosed tubes.

The mass flow rate of a fluid flowing in a tube is given by the product of the fluid density ρ, the mean velocity U and the cross-sectional fluid flow area A_c (where suffix c distinguishes this area from the heat flow area A). At a section of the tube denoted by suffix 1

$$\dot{m}_1 = \rho_1 U_1 A_{c1}$$

and at another section denoted by suffix 2

$$\dot{m}_2 = \rho_2 U_2 A_{c2}$$

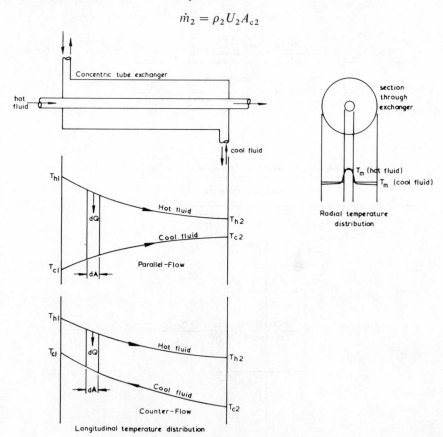

Figure 5.2 Temperature distributions in concentric tube exchangers.

Under conditions of steady flow in a continuous tube the mass flow rate is constant along the tube and

$$\dot{m}_1 = \dot{m}_2$$

These expressions are referred to as the continuity equations for steady flow along a continuous tube.

5.2 Average Temperatures

In this and the following sections the attention of the reader will be directed towards the analysis of a simple concentric tube exchanger of the type shown in Fig. 5.2. The object is to determine the relationships between the enthalpy exchange and the temperature differences involved. Complication arises because the fluid temperatures vary in both the radial and longitudinal directions. Average temperatures are therefore specified in both directions such that the overall heat transfer may be expressed in the form $Q = UA\,\Delta T_\text{m}$ where ΔT_m is a suitable mean temperature difference.

In the radial direction the mean temperature at any cross-section of flow through a tube is conventionally defined as that fluid temperature which would result if it was thoroughly mixed at that section. This average value is termed the mean mixed or bulk fluid temperature T_m and is defined in terms of the velocity u at any point across the tube section as:

$$\dot{m}cT_\text{m} = (\rho U A_\text{c})cT_\text{m} = \rho c \int_{A_\text{c}} uT\,\mathrm{d}A_\text{c}$$

i.e.

$$T_\text{m} = \frac{\displaystyle\int_{A_\text{c}} uT\,\mathrm{d}A_\text{c}}{UA_\text{c}} \tag{5.2}$$

where A_c is the cross-sectional area of flow and U is the bulk velocity defined by the continuity equation, (see also equation (8.6)). Mean convection heat transfer coefficients are conventionally based on T_m. This point is discussed further in Chapter 8, but in general, when a mean temperature is allotted to a fluid flowing in a duct, it is taken to be the bulk fluid temperature (and the suffix m is normally omitted to avoid over-complicating the notation).

In the longitudinal direction the mean temperature difference ΔT_m for the parallel flow case shown in Fig. 5.2 may be determined as follows. The steady heat flow rate $\mathrm{d}Q$ transferred in a section of the exchanger $\mathrm{d}A$ is obtained from equation (5.1) (with positive heat flow from the hotter to the cooler fluid):

$$\mathrm{d}Q = -\dot{m}_\text{h}c_\text{h}\,\mathrm{d}T_\text{h} = \dot{m}_\text{c}c_\text{c}\,\mathrm{d}T_\text{c}$$

and therefore

$$dT_h = \frac{-dQ}{\dot{m}_h c_h} \quad \text{and} \quad dT_c = \frac{dQ}{\dot{m}_c c_c}$$

Also

$$d(T_h - T_c) = dT_h - dT_c$$

$$= -dQ\left(\frac{1}{\dot{m}_h c_h} + \frac{1}{\dot{m}_c c_c}\right) \tag{5.3}$$

The heat transfer rate in the increment may also be expressed in terms of the overall heat transfer coefficient U as

$$dQ = U(T_h - T_c)\,dA \tag{5.4}$$

where T_h and T_c refer to bulk fluid temperatures as discussed previously. Elimination of dQ between these two equations leads to

$$\frac{d(T_h - T_c)}{(T_h - T_c)} = -U\left(\frac{1}{\dot{m}_h c_h} + \frac{1}{\dot{m}_c c_c}\right)dA$$

Integration over the length of the exchanger between limits of 1 and 2 as defined in Fig. 5.2 gives

$$\ln\left(\frac{T_{h2} - T_{c2}}{T_{h1} - T_{c1}}\right) = -UA\left(\frac{1}{\dot{m}_h c_h} + \frac{1}{\dot{m}_c c_c}\right) \tag{5.5}$$

Substitution of

$$\dot{m}_h c_h = \frac{-Q}{T_{h2} - T_{h1}} \tag{5.6}$$

and

$$\dot{m}_c c_c = \frac{Q}{T_{c2} - T_{c1}} \tag{5.7}$$

then yields:

$$Q = UA\,\frac{(T_{h2} - T_{c2}) - (T_{h1} - T_{c1})}{\ln[(T_{h2} - T_{c2})/(T_{h1} - T_{c1})]} \tag{5.8}$$

Comparison with the following expression for the total heat transfer rate for the exchanger in terms of a mean temperature,

$$Q = UA\,\Delta T_m \tag{5.9}$$

gives

$$\Delta T_m = \frac{\Delta T_2 - \Delta T_1}{\ln(\Delta T_2/\Delta T_1)} \tag{5.10}$$

where $\Delta T_2 = T_{h2} - T_{c2}$, $\Delta T_1 = T_{h1} - T_{c1}$ and the denominator is a natural logarithm. For exchangers equation (5.9) has no negative sign on the right-hand side and ΔT_m is always positive, leading to a positive heat flow from the hotter to the cooler fluid.

The term ΔT_m is called the log mean temperature difference and it is found to be given by the same expression for counter-flow exchangers. Its derivation involves the assumption that the specific heats of both fluids are constant with temperature change and that the heat transfer coefficients are constant along the length of the exchanger. In the case of a counter-flow exchanger with $\Delta T_1 = \Delta T_2$ the expression for ΔT_m yields an indeterminate value but in this situation the fluid temperature distributions are parallel and obviously $\Delta T_m = \Delta T_1 = \Delta T_2$. The log mean temperature difference approach is summarized in Fig. 5.3 and some examples of its use are given in Section 5.4.

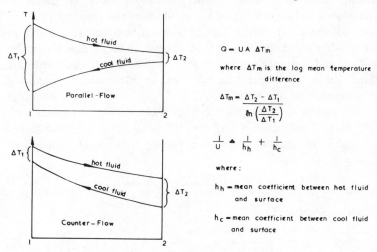

$$Q = UA\ \Delta T_m$$

where ΔT_m is the log mean temperature difference

$$\Delta T_m = \frac{\Delta T_2 - \Delta T_1}{\ln\left(\frac{\Delta T_2}{\Delta T_1}\right)}$$

$$\frac{1}{U} = \frac{1}{h_h} + \frac{1}{h_c}$$

where:

$h_h =$ mean coefficient between hot fluid and surface

$h_c =$ mean coefficient between cool fluid and surface

Figure 5.3   The log mean temperature difference.

5.3   Exchanger Effectiveness

The analysis of exchangers using the log mean temperature difference and leading to equations (5.9), and (5.10) is useful for determining the heat transfer coefficients of existing exchangers where the fluid inlet and exit conditions can be measured. The design of exchangers is generally based on known fluid inlet conditions and estimated heat transfer coefficients. The unknown parameters are then the outlet conditions and the heat transfer or the surface area required for a given heat transfer. The outlet condition for each fluid may be found by eliminating Q between equations (5.6), (5.7), and (5.8) but the resulting equations require a trial-and-error solution owing to the log term. A more convenient exchanger analysis is offered by an approach based on the concepts of capacity ratio, effectiveness and number of transfer units. In addition, this approach facilitates comparison between the various types of exchangers which may be used for a particular application.

The product $\dot{m}c$ of a fluid flowing in an exchanger is repeatedly encountered in the theory and is termed the capacity rate as it indicates the capacity of the fluid to store energy at a given rate. The capacity rate ratio C (or simply the capacity ratio) is defined as the ratio of the minimum to the maximum capacity rate:

$$C = \frac{(\dot{m}c)_{min}}{(\dot{m}c)_{max}} \tag{5.11}$$

In parallel-flow or counter-flow exchangers the hotter or cooler fluid may have the minimum $\dot{m}c$ value.

The effectiveness is defined as the ratio of the energy actually transferred to the maximum theoretical energy transfer. The actual energy transferred is given by equations (5.6) and (5.7) as the product of the capacity rate and temperature difference for either fluid. The maximum theoretical energy transfer occurs when the temperature difference between the fluids is maximum. In parallel-flow and counter-flow exchangers the fluid exit temperatures are both situated within the temperature range of the inlet conditions and the maximum difference is therefore $(T_{h\ inlet} - T_{c\ inlet})$. If one of the fluids underwent this maximum temperature change it could only be the one with the smaller value of $\dot{m}c$, since $\dot{m}c\ \Delta T$ has the same value for both fluids, (equation (5.1)). The maximum possible energy transfer Q_{max} therefore becomes

$$Q_{max} = (\dot{m}c)_{min}(T_{h\ inlet} - T_{c\ inlet}) \tag{5.12}$$

The effectiveness E is then

$$E = \frac{(\dot{m}c\ \Delta T)_{fluid}}{(\dot{m}c)_{min}(T_{h,\ inlet} - T_{c,\ inlet})} \tag{5.13}$$

where the numerator refers to either fluid and is always positive. For any particular exchanger E may be expressed solely in terms of the temperatures. For example, the effectiveness E_{ph} in the case of a parallel-flow exchanger with the hotter fluid having minimum capacity rate becomes

$$E_{ph} = \frac{(\dot{m}c)_h(\Delta T)_h}{(\dot{m}c)_h(T_{h1} - T_{c1})} = \frac{T_{h1} - T_{h2}}{T_{h1} - T_{c1}} \tag{5.14}$$

Consideration of the other possibilities indicates that the effectiveness may be memorized as shown in Fig. 5.4 for the cases of parallel-flow and counter-flow.

A further parameter encountered in this approach to heat exchangers is termed the Number of Transfer Units, *NTU*, defined as

$$NTU = \frac{UA}{(\dot{m}c)_{min}} \tag{5.15}$$

This term provides some indication of the physical size of a heat exchanger. It is now proposed to examine a parallel-flow exchanger using the effectiveness approach and derive the relationship between the terms C, E and *NTU*.

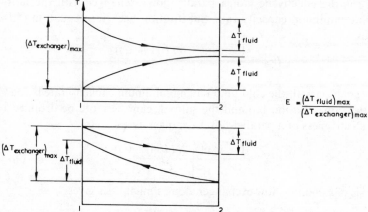

Figure 5.4 Effectiveness E in terms of temperatures.

From equation (5.3) we have

$$d(T_h - T_c) = -dQ\left(\frac{1}{(\dot{m}c)_h} + \frac{1}{(\dot{m}c)_c}\right)$$

When the hot fluid has minimum capacity rate so that $C = (\dot{m}c)_h/(\dot{m}c)_c$ this becomes

$$(\dot{m}c)_h \, d(T_h - T_c) = -dQ(1 + C)$$

Substitution of equation (5.4) for dQ and rearrangement yields

$$\frac{d(T_h - T_c)}{(T_h - T_c)} = \frac{-U \, dA}{(\dot{m}c)_h}(1 + C)$$

Integration and use of equation (5.15) then gives

$$\frac{T_{h2} - T_{c2}}{T_{h1} - T_{c1}} = \exp(-NTU(1 + C)) \tag{5.16}$$

This equation may be rearranged by using equation (5.1) in the form

$$C = \frac{\dot{m}_h c_h}{\dot{m}_c c_c} = \frac{T_{c2} - T_{c1}}{T_{h1} - T_{h2}}$$

from which

$$T_{c2} = C(T_{h1} - T_{h2}) + T_{c1}$$

and the left-hand side becomes

$$\frac{T_{h2} - T_{c2}}{T_{h1} - T_{c1}} = \frac{T_{h2} - T_{c1} - C(T_{h1} - T_{h2})}{T_{h1} - T_{c1}}$$

$$= \frac{-(T_{h1} - T_{h2}) + (T_{h1} - T_{c1}) - C(T_{h1} - T_{h2})}{T_{h1} - T_{c1}}$$

$$= -E_{ph} + 1 - CE_{ph}$$

$$= 1 - E_{ph}(1 + C)$$

where E_{ph} is the effectiveness of a parallel-flow exchanger with the hotter fluid having minimum capacity rate. Substitution back into equation (5.16) then yields

$$E_{ph} = \frac{1 - \exp(-NTU(1 + C))}{1 + C}$$

A similar analysis with the cold fluid having minimum capacity rate is found to yield the same relationship and the suffix h can therefore be dropped to give the effectiveness of a parallel-flow exchanger as

$$E_p = \frac{1 - \exp(-NTU(1 + C))}{1 + C} \tag{5.17}$$

An analysis of a counter-flow exchanger along similar lines gives

$$E_c = \frac{1 - \exp(-NTU(1 - C))}{1 - C \exp(-NTU(1 - C))} \tag{5.18}$$

The analysis of other types of exchangers is more complicated owing to the possibility of mixed or unmixed fluid arrangements and two- or three-dimensional flow. A correction factor F is generally applied in the case of the log mean temperature difference based approach such that

$$Q = UAF \, \Delta T_m \tag{5.19}$$

and charts of F for various geometrical and thermal conditions are available in the literature. However, the effectiveness approach using charts of E as a function of NTU and C is more commonly used for design purposes, (see Kays and London, 1964).

5.4 Exchanger Examples

Before we evaluate numerically one or two examples of exchanger problems there are a few general points regarding exchanger calculations which should be noted. No specific mention has been made of boilers and condensers although they form one of the most important groups of exchangers. Basically they may be evaluated in the same way as the previous exchangers we have considered, although the fact that one fluid undergoes no temperature change leads to some simplification. In particular C is found to be zero and the effectiveness is given by $E = 1 - \exp(-NTU)$. The area on which the overall heat transfer coefficient is based (for all tubular exchangers) is generally that value obtained by using the outer diameter of the exchanger tubes. It is usually acceptable to ignore the thermal resistance of the metal wall dividing the fluids and the overall heat transfer coefficient is therefore given by the fluid to wall coefficients as indicated in Fig. 5.3. The energy loss to the surroundings is assumed to be negligible compared to the energy exchange.

Example 5.1. The heat energy for a space heating system is extracted from the exhaust gases of an internal combustion engine. The single-pass, multi-tube heat exchanger employed operates in counter-flow and has a nominal tube diameter of 20 mm. The exhaust gases with a mean specific heat of 1.1 kJ/kg K and mean gas constant R of 0.3 kJ/kg K flow through the tubes with a velocity of 20 m/s and are cooled from 300 °C to 100 °C. Water flowing out-side the tubes is heated from 20 °C to 90 °C and the overall heat transfer coefficient may be taken as 0.2 kW/m² K.

These conditions apply when the internal combustion engine is producing a brake power of 100 kW. The specific fuel consumption is 0.2 kg/kW h, the air-fuel ratio is 16 to 1 and the exhaust gas pressure is 100 mm water. An estimate is required of the number and length of the exchanger tubes and the mass flow rate of water.

Solution. The number of tubes may be determined from the continuity equation in the form:

$$\dot{m}_g v = V A_c$$

The mass flow rate of the gases $\dot{m}_g$ may be found from the engine data:

$$\text{fuel flow} = 0.2 \times 100 = 20 \text{ kg/h}$$
$$\text{air flow} = 16 \times \text{fuel flow} = 320 \text{ kg/h}$$
$$\text{total exhaust gas flow } \dot{m}_g = 340 \text{ kg/h}$$

The specific volume $v(=1/\rho)$ may be estimated from the perfect gas relation-ship at the exhaust temperature of 300 °C and pressure of 1 atmosphere plus 100 mm water head:

$$v = \frac{RT}{p} = \frac{0.3 \times (300 + 273) \times 10^3}{(1.013 + 0.01) \times 10^5}$$
$$= 1.68 \text{ m}^3/\text{kg}$$

By substitution in the continuity equation the total cross-sectional area A_c becomes

$$A_c = \frac{\dot{m}_g v}{V} = \frac{340 \times 1.68}{3600 \times 20}$$
$$= 7.94 \times 10^{-3} \text{ m}^2$$

Division of this value by the cross-sectional area per tube yields the number n of tubes required:

$$n = \frac{A_c}{A_{tube}} = \frac{7.94 \times 4 \times 10^{-3}}{\pi \times 4 \times 10^{-4}} = 25.3$$

$$\text{say } n = 26 \text{ tubes}$$

The energy transferred from the gases per unit time Q is given by

$$Q = \dot{m}_g c_g \, \Delta T_g$$

$$= \frac{340}{3600} \times 1.1 \times 200 = 20.8 \text{ kW}$$

The water mass flow rate is given by

$$\dot{m}_w = \frac{Q}{c_w \, \Delta T_w} = \frac{20.8}{4.18 \times 70}$$

$$\underline{\dot{m}_w = 0.711 \text{ kg/s}}$$

The exchanger overall heat transfer relationship is

$$Q = UA \, \Delta T_m$$

where A is the heat transfer surface area and

$$\Delta T_m = \frac{\Delta T_2 - \Delta T_1}{\ln(\Delta T_2 / \Delta T_1)}$$

Substitution of the temperatures (as illustrated in Fig. 5.5) gives

$$\Delta T_m = \frac{(100 - 20) - (300 - 90)}{\ln(80/210)}$$

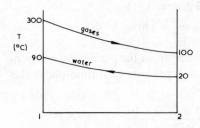

Figure 5.5

or more conveniently for calculation

$$\Delta T_m = \frac{210 - 80}{\ln(210/80)} = 134.6 \,°\text{C}$$

Thus:

$$A = \frac{Q}{U \, \Delta T_m} = \frac{20.8}{0.2 \times 134.6} = 0.773 \text{ m}^2$$

and the length of the tubes is

$$l = \frac{A}{\pi dn} = \frac{0.773}{\pi \times 0.02 \times 26}$$

$$\underline{l = 0.472 \text{ m}}$$

80

Example 5.2. Oil with a specific heat of 2.0 kJ/kg K is cooled from 110 °C to 75 °C by a flow of water in a parallel-flow exchanger. The water flows at the rate of 70 kg/min and is heated from 35 to 70 °C. The overall heat transfer coefficient is estimated to be 0.32 kW/m² K. It is required to find the effect on the exit temperatures of the oil and water if the water flow rate drops to 50 kg/min at the same oil flow rate.

Solution. The heat exchange area may be found from

$$Q = UA \, \Delta T_m$$

where

$$Q = \dot{m}_w c_w \, \Delta T_w$$

$$= \frac{70}{60} \times 4.18 \times 35 = 171 \text{ kJ/s}$$

$$\Delta T_m = \frac{75 - 5}{\ln(75/5)} = 25.9 \text{ °C (see Fig. 5.6)}$$

and

$$A = \frac{171}{0.32 \times 25.9} = 20.6 \text{ m}^2$$

The oil flow rate is

$$\dot{m}_0 = \frac{Q}{c_0 \, \Delta T_0} = \frac{171}{2.0 \times 35} = 2.44 \text{ kg/s}$$

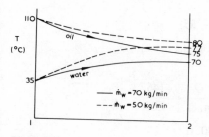

Figure 5.6

The effectiveness approach is now used to determine the temperature when $\dot{m}_w$ is reduced from 70 to 50 kg/min.
Capacity rates are:

$$\dot{m}_0 c_0 = 2.44 \times 2.0 = 4.88$$

$$\dot{m}_w c_w = \frac{50}{60} \times 4.18 = 3.48$$

Therefore

$$(\dot{m}c)_{min} = 3.48$$

and
$$C = \frac{(\dot{m}c)_{min}}{(\dot{m}c)_{max}} = 0.713$$

$$NTU = \frac{UA}{(\dot{m}c)_{min}} = \frac{0.32 \times 20.6}{3.48}$$

$$= 1.89$$

From equation (5.17) giving the effectiveness E_p of a double-pipe exchanger:
$$E_p = \frac{1 - \exp(-NTU(1 + C))}{1 + C}$$

and substitution yields
$$E_p = 0.561$$

Also
$$E_p = \frac{T_{w2} - T_{w1}}{T_{01} - T_{w1}}$$

i.e.,

$$0.561 = \frac{T_{w2} - 35}{110 - 35}$$

$$T_{w2} = 77\,^{\circ}C$$

The new heat flow rate is
$$Q = \dot{m}_w c_w \, \Delta T_w$$

$$= 3.48 \times (77 - 35)$$

$$= 146 \text{ kJ/s (note decrease)}$$

and the oil exit temperature is
$$110 - T_{02} = \frac{Q}{\dot{m}_0 c_0} = \frac{146}{4.88}$$

$$T_{02} = 80\,^{\circ}C$$

The usefulness of the effectiveness approach may be gauged if this example is attempted using the log mean temperature approach alone.

Example 5.3. A single-pass steam condenser contains one hundred thin-walled tubes of 25 mm nominal diameter and 2 m length. Cooling water enters at a temperature of 10 °C, leaves at 50 °C and flows through the tubes at a velocity of 2 m/s. The condenser pressure is 0.5 bar and the condensing heat transfer coefficient is 5 kW/m² K. It is required to determine the condensate flow rate and the mean temperature of the tube metalwork at the centre of the condenser length.

Solution. The heat transfer coefficient on the cooling water side may be determined from the Dittus–Boelter relationship (equation (3.11) or Table 3.2):

$$Nu = 0.023 \ Re^{0.8} \ Pr^{0.4}$$

with properties from tables (such as Mayhew and Rogers, 1969) at a mean temperature of, say, 30 °C. Thus:

$$Re = \frac{\rho V d}{\mu}$$

$$= \frac{1000 \ (kg/m^3) \times 2(m/s) \times 25 \times 10^{-3} \ (m)}{797 \times 10^{-6} \ (kg/m \ s)}$$

$$= 62,700$$

$$Pr = \frac{c_p \mu}{k}$$

$$= \frac{4.179 \ (kJ/kg \ K) \times 797 \times 10^{-6} \ (kg/m \ s)}{0.618 \times 10^{-3} \ (kW/m \ K)}$$

$$= 5.39$$

and

$$Nu = 0.023 \times (62,700)^{0.8} \times (5.39)^{0.4}$$

$$= 312$$

From

$$Nu = \frac{hd}{k}$$

$$h = \frac{312 \times 0.618 \times 10^{-3} \ (kW/m \ K)}{25 \times 10^{-3} \ (m)}$$

$$= 7.71 \ kW/m^2 \ K$$

If the thermal resistance of the tube is negligible, the overall heat transfer co-efficient is given by the following expression, (where suffixes o and i indicate outer and inner):

$$\frac{1}{U} = \frac{1}{h_0} + \frac{1}{h_i}$$

$$= \frac{1}{5} + \frac{1}{7.71}$$

$$U = 3.04 \ kW/m^2 \ K$$

At a pressure of 0.5 bar the condensing temperature is 81.3 °C from tables. Ignoring the condensate subcooling the log mean temperature difference becomes (see Fig. 5.7):

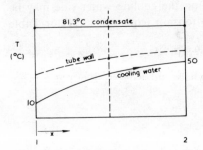

Figure 5.7

$$\Delta T_m = \frac{\Delta T_1 - \Delta T_2}{\ln(\Delta T_1/\Delta T_2)}$$

$$= \frac{(81.3 - 10) - (81.3 - 50)}{\ln(71.3)/(31.3)}$$

$$= 48.6\,°\text{C}$$

and

$$Q = UA\,\Delta T_m$$

$$= 3.04 \times \pi(25 \times 10^{-3})2 \times 48.6$$

$$= 23.2\,\text{kW/tube}$$

The mass flow rate of condensate $\dot{m}_c$ is obtained from

$$Q = \dot{m}_c h_{fg}$$

where h_{fg} is the enthalpy of vaporization (or latent heat). For 100 tubes

$$\dot{m}_c = \frac{100 \times 23.2}{2305}$$

$$\dot{m}_c \approx 1\,\text{kg/s}$$

In order to find the tube metalwork temperature at the centre of the exchanger it is first necessary to calculate the cooling water temperature at this point. The general equation for the temperature difference between the fluids may be expressed as

$$\log_{10} \Delta T = Ax + B$$

where x is the distance along the exchanger as shown in Fig. 5.7. With the conditions

$$\Delta T = 81.3 - 10 = 71.3 \text{ at } x = 0$$

$$\Delta T = 81.3 - 50 = 31.3 \text{ at } x = 2 \text{ m}$$

the equation becomes

$$\log_{10} \Delta T = -0.174\,x + 1.853$$

84

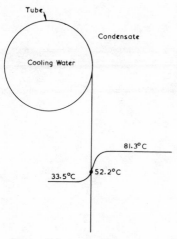

81.3°C

33.5°C 52.2°C

Figure 5.8

and at $x = 1$ m

$$\Delta T = 47.8$$

and $\qquad T_c = 81.3 - 47.8 = 33.5\,°C$ at the midpoint.

The radial conditions at the midpoint are shown in Fig. 5.8 and the radial heat flow is given by

$$Q = h_i A(T_w - 33.5) = h_o A(81.3 - T_w)$$

Substitution of h_i and h_o then yields a value for the wall temperature T_w of

$$\underline{T_w = 52.2\,°C \text{ at the centre}}$$

An Introduction to the Analysis of Heat Flow

Radiation Analysis

6

6.1 Simple Arrangements involving Black Surfaces

The basic relationship for radiation heat transfer between two black surfaces is given by equation (1.10) as

$$Q_{12} = -A_1 F_{12} \sigma (T_2^4 - T_1^4) \tag{6.1}$$

where F_{12}, the shape factor, is the fraction of the total radiation from area A_1 which falls on area A_2. The calculation of the shape factor necessitates integration of the radiation intensity over the two areas involved as shown in Section 1.4 and is generally not possible except for the simplest geometries. Methods of determining the shape factor will be examined later in the chapter, and in this section we shall consider one or two cases where the value is self-evident from its definition as the fraction of the total radiation from a surface.

Figure 6.1 shows two balls which we shall assume are situated a great

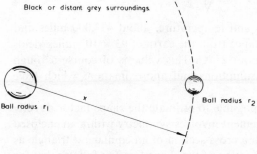

Black or distant grey surroundings

Ball radius r_1

x

Ball radius r_2

Figure 6.1 Radiation between black spheres.

distance apart in surroundings which are effectively black. (This includes surroundings which are non-black but at a considerable distance from the balls so that virtually none of the radiation from the balls is reflected back to them from the surroundings. Bench top experiments conducted in a laboratory, for example, are generally assumed to have black surroundings.)

The radiation from ball 1 spreads out evenly in all directions and we shall imagine that it falls on the inside of a hollow sphere of radius x. Since the radiation is spread evenly on the hollow sphere it follows that the proportion of the total radiation falling on body 2 from body 1 is the ratio of the projected area of body 2 to the total area of the sphere. In this case therefore

$$F_{12} = \frac{\pi r_2^2}{4\pi x^2}$$

where r_2 is the radius of ball 2. Substitution in equation (6.1) yields

$$Q_{12} = -\frac{\pi r_1^2 r_2^2}{x^2} \sigma(T_2^4 - T_1^4)$$

where r_1 is the radius of ball 1. This then yields an expression for the radiation from ball 1 to 2 or, since the expression is symmetrical with respect to the balls, from ball 2 to 1. It does not of course account for the radiation transfer between the balls and the surroundings and problems of this type are solved in Section 6.4. One case in which this expression does yield the complete radiation transfer is when the surroundings do not emit any radiation, that is when the temperature of the surroundings is $0\,°K$. An example of this is the solar system where, in our case, ball 1 becomes the sun and ball 2 the earth or some other planet. The radiation emitted from the earth to outer space is given by

$$Q_{20} = -4\pi r_2^2 \times 1 \times \sigma(0 - T_2^4)$$

Assuming the heat received from the sun is equal to the heat radiated from the earth, the right-hand sides of the last two equations may be equated to yield, after neglecting a small term;

$$T_2 = T_1 \sqrt{\frac{r_1}{2x}}$$

Substitution of the sun's radius and temperature, about 433,000 miles and $6200\,°K$ respectively, and its distance from the earth of 93×10^6 miles yields a mean earth temperature of about $27\,°C$. This value is of course a rough estimate owing to the many assumptions and approximations which have been made.

Another situation where it is possible to estimate the shape factor directly from the geometry of the arrangement involves symmetry within an enclosed space. Consider a long tube with a cross-section of an equilateral triangle as shown in Fig. 6.2a. From the definition of the shape factor it follows that for

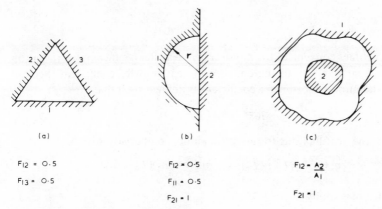

(a)

$F_{12} = 0.5$

$F_{13} = 0.5$

(b)

$F_{12} = 0.5$

$F_{11} = 0.5$

$F_{21} = 1$

(c)

$F_{12} = \dfrac{A_2}{A_1}$

$F_{21} = 1$

Figure 6.2 Shape factors for enclosed spaces.

radiation from surface 1, $F_{12} + F_{13} = 1$ and since the radiation is equally divided between surfaces 2 and 3 it follows that $F_{12} = F_{13} = 0.5$. A less obvious case is shown in Fig. 6.2b where radiation is between a hemispherical surface 1 and a plane surface 2. In this situation the hemispherical surface can 'see' itself and the shape factor for the proportion of total radiation which falls on itself is denoted by F_{11}. Since the rest of the radiation falls on surface 2 it follows that

$$F_{11} + F_{12} = 1$$

From the reciprocity relationship, equation (1.17),

$$A_1 F_{12} = A_2 F_{21}$$

Furthermore $F_{21} = 1$ and therefore

$$F_{12} = \frac{A_2}{A_1}$$

$$= \frac{\pi r^2}{4\pi r^2 / 2}$$

$$= 0.5$$

and

$$F_{11} = 0.5$$

Thus half of the radiation from the hemisphere falls on itself and half on surface 2. In the case shown in Fig. 6.2c it follows that

$$F_{21} = 1$$

$$F_{12} + F_{11} = 1$$

and as

$$A_1 F_{12} = A_2 F_{21}$$

$$F_{12} = \frac{A_2}{A_1}$$

and
$$F_{11} = 1 - \frac{A_2}{A_1}$$

It is sometimes convenient to employ a heat transfer coefficient for radiation defined in a similar way to the convection heat transfer coefficient (Section 2.2). If a radiation heat transfer coefficient is defined such that

$$q_r = -h_r(T_2 - T_1)$$

where q_r is the heat flux due to radiation alone, comparison with

$$q_r = -\sigma(T_2^4 - T_1^4) = -\sigma(T_2^2 - T_1^2)(T_2^2 + T_1^2)$$

yields for a blackbody

$$h_r = \sigma(T_2 + T_1)(T_2^2 + T_1^2) \qquad (6.2)$$

In cases when one temperature, say T_1, is much greater than T_2 as in radiation to ambient surroundings from a furnace, an electric arc or an electric fire it is found that h_r is approximately proportional to T_1^3.

6.2 Parallel Grey Surfaces and Radiation Shields

In the previous section we examined the transfer of heat by radiation between simple arrangements of black surfaces and therefore avoided the complication of the radiation surface resistance. In this section it is proposed to concentrate on the radiation between grey surfaces and avoid the complication of the space resistance by considering situations where the shape factor is unity. After a more detailed examination of the shape factor in Section 6.3 we shall attempt to combine these areas in Section 6.4 so that engineering problems involving both space and surface resistances may be solved.

Consider two large parallel grey surfaces denoted 1 and 2. (The following reasoning may also be applied to curved surfaces where the radius of curvature is large compared to the gap between the surfaces.)

Radiation emitted by surface 1 $= \varepsilon_1 A\sigma T_1^4$

Radiation absorbed by surface 2 $= \alpha_2(\varepsilon_1 A\sigma T_1^4)$

$$= \varepsilon_1 \varepsilon_2 A\sigma T_1^4$$

Radiation reflected by surface 2 $= \rho_2(\varepsilon_1 A\sigma T_1^4)$

Of this radiation reflected by surface 2:

Radiation absorbed by surface 1 $= \alpha_1(\rho_2\varepsilon_1 A\sigma T_1^4)$

Radiation reflected by surface 1 $= \rho_1(\rho_2\varepsilon_1 A\sigma T_1^4)$

Of this radiation reflected by surface 1:

Radiation absorbed by surface $2 = \alpha_2 \rho_1 (\rho_2 \varepsilon_1 A \sigma T_1^4)$

$$= \rho_1 \rho_2 \varepsilon_1 \varepsilon_2 A \sigma T_1^4$$

Continuation of this process shows that the total energy absorbed by surface 2 is

$$Q_2 = \varepsilon_1 \varepsilon_2 A \sigma T_1^4 [1 + \rho_1 \rho_2 + (\rho_1 \rho_2)^2 + (\rho_1 \rho_2)^3 \ldots]$$

Similarly the total energy absorbed by surface 1 is

$$Q_1 = \varepsilon_1 \varepsilon_2 A \sigma T_2^4 [1 + \rho_1 \rho_2 + (\rho_1 \rho_2)^2 + (\rho_1 \rho_2)^3 \ldots]$$

The energy exchange is given by

$$Q_{12} = Q_1 - Q_2$$

Substitution and summation of the converging series yields

$$Q_{12} = \frac{\varepsilon_1 \varepsilon_2 A \sigma (T_1^4 - T_2^4)}{1 - \rho_1 \rho_2}$$

Substitution of $\rho_1 = 1 - \alpha_1 = 1 - \varepsilon_1$ and $\rho_2 = 1 - \varepsilon_2$ and rearrangement yields

$$Q_{12} = \frac{- \sigma A (T_2^4 - T_1^4)}{1/\varepsilon_1 + 1/\varepsilon_2 - 1} \qquad (6.3)$$

This expression may also be obtained using the network analogy discussed in Section 1.4. In the case of two large parallel surfaces as shown in Fig. 1.10 the heat flow between the surfaces becomes, on summing the resistances:

$$Q_{12} = \frac{\sigma T_1^4 - \sigma T_2^4}{\dfrac{1 - \varepsilon_2}{A_2 \varepsilon_2} + \dfrac{1}{A_1 F_{12}} + \dfrac{1 - \varepsilon_1}{\varepsilon_1 A_1}}$$

In this case $A_1 = A_2 = A$ and $F_{12} = 1$ so that

$$Q_{12} = \frac{- A \sigma (T_2^4 - T_1^4)}{\dfrac{1 - \varepsilon_2}{\varepsilon_2} + 1 + \dfrac{1 - \varepsilon_1}{\varepsilon_1}}$$

and rearrangement gives equation (6.3).

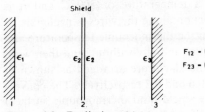

Figure 6.3 Radiation shield.

Radiation shields are sometimes used to reduce the heat transfer between surfaces by effectively increasing the number of surface resistances. In the system shown in Fig. 6.3 the heat flow to the shield from surface 1 is equal to the flow to surface 3 and from equation (6.3)

$$Q = \frac{-A\sigma(T_2^4 - T_1^4)}{1/\varepsilon_1 + 1/\varepsilon_2 - 1} = \frac{-A\sigma(T_3^4 - T_2^4)}{1/\varepsilon_2 + 1/\varepsilon_3 - 1}$$

where the emissivity ε_2 both sides of the shield is the same. If the shield surfaces were perfectly reflective ($\varepsilon_2 = 0$) there would be no heat transfer and in practice they generally have polished metal surfaces in order to reduce ε_2 as much as possible. When all the surfaces have the same emissitivity

$$T_2^4 = \tfrac{1}{2}(T_3^4 + T_1^4),$$

and on substitution

$$Q = \frac{-\tfrac{1}{2}\sigma A(T_3^4 - T_1^4)}{1/\varepsilon + 1/\varepsilon - 1}$$

and the direct radiation heat transfer between surfaces 1 and 3 is halved by the shield. More than one shield is sometimes used and in the general case where there are n shields (with all the shield and surface emissivities equal) the sum of the surface resistances and space resistances per unit area is

$$R_n = (2n + 2)\left(\frac{1 - \varepsilon}{\varepsilon}\right) + (n + 1) \times 1 = (n + 1)\left(\frac{2}{\varepsilon} - 1\right)$$

With no shields ($n = 0$) the resistance is

$$R_0 = \frac{2}{\varepsilon} - 1$$

and the heat transfer with shields Q_n may be related to the heat transfer without shields Q_0 by

$$\frac{Q_n}{Q_0} = \frac{R_0}{R_n} = \frac{1}{n + 1}$$

assuming of course that the reduction in heat flow does not affect the wall temperatures.

Example 6.1. A thermocouple is used to measure the temperature of combustion gases flowing into a room with an ambient room temperature T_r of 20 °C. The thermocouple indicates a temperature of 500 °C and it is arranged so that heat transfer by conduction along the wires is negligible. It is required to estimate the error between the thermocouple temperature and the gas stream temperature due to radiation initially without, and then with, a radiation shield. The thermocouple and shield are arranged as shown in Fig. 6.4 and the surface emissivities are 0.6 and 0.3 respectively. The convective heat transfer coefficient between the shield (and thermocouple surface) and the gas stream may be taken as 0.2 kW/m² K.

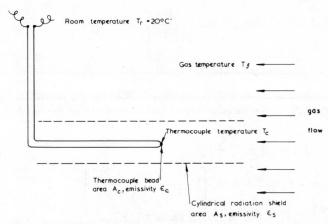

Figure 6.4 Thermocouple in a gas stream.

Solution. In the case with no radiation shield there exists an equilibrium between the heat flow to the thermocouple by convection and heat flow from the thermocouple to the surroundings by radiation. The surroundings which are non-black but distant are effectively black, as negligible radiation is reflected back to the thermocouple bead. In the absence of conduction the balance between convection and radiation yields

$$-hA_c(T_f - T_c) = \varepsilon_c A_c \sigma(T_r^4 - T_c^4)$$

where the notation is as shown in Fig. 6.4. Substitution of numerical values gives

$$-0.2(T_f - 773) = 0.6 \times 56.7 \left[\left(\frac{293}{1000} \right)^4 - \left(\frac{773}{1000} \right)^4 \right]$$

$$(T_f - 773) = 59.6\,^\circ\text{K}$$

and the error is therefore $59.6\,^\circ$K.

When the radiation shield is in place, radiation from the thermocouple is determined by the temperature of the shield. If the cylindrical shield is long compared to its diameter a reasonable approximation is that radiation from the shield to the surroundings is entirely from the external surface and that radiation from the thermocouple bead is entirely to the inner surface. Furthermore radiation from the thermocouple bead to the radiation shield is assumed to be negligible compared to radiation between the shield and the surroundings (in view of the surface areas and temperature differences involved). Under these conditions a balance between the convective and radiative heat transfer to the shield of area A_s yields

$$-2hA_s(T_f - T_s) = \varepsilon_s A_s \sigma(T_r^4 - T_s^4)$$

where the factor 2 arises because convection occurs both outside and inside the shield. If $T_f = 773 + 59.6 = 832.6\,^\circ$K numerical substitution yields

95

$$-2 \times 0.2(832.6 - T_s) = 0.3 \times 56.7\left[\left(\frac{293}{1000}\right)^4 - \left(\frac{T_s}{1000}\right)^4\right]$$

and trial and error solution gives

$$T_s = 814\,^\circ\text{K}$$

A balance between radiation from the thermocouple bead to the shield (a distant greybody compared to the thermocouple) and convection to the bead yields

$$-hA_c(T_f - T_c) = \varepsilon_c A_c \sigma(T_s^4 - T_c^4)$$

$$-0.2(832.6 - T_c) = 0.6 \times 56.7\left[\left(\frac{814}{1000}\right)^4 - \left(\frac{T_c}{1000}\right)^4\right]$$

$$T_c = 828\,^\circ\text{K} \text{ (to the nearest } 1\,^\circ\text{K)}$$

The error with a shield is therefore about $5\,^\circ$K compared to $60\,^\circ$K with no shield. This error could be further reduced by a more reflective shield with a lower ε_s value.

6.3   Shape Factor Determination

Now that we have studied the radiation between surfaces arranged in simple configurations we shall proceed to cases of a more complex nature. In this section we shall once again consider entirely black surfaces and therefore direct attention to the geometrical configuration of the surfaces and the shape

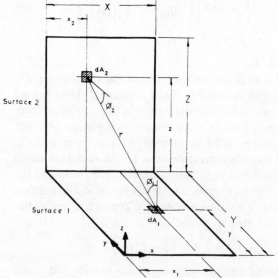

Figure 6.5   Co-ordinate system for determining the shape factor.

factor. Analytically the problem involves performing the double area integration as indicated in equation (1.16):

$$A_1 F_{12} = \int_{A_1} \int_{A_2} \frac{\cos \phi_1 \cos \phi_2}{\pi r^2} \, dA_1 \, dA_2$$

For surface shapes and orientations which can be simply represented in Cartesian or radial coordinates it is possible to perform the integration mathematically but in other situations it may be necessary to use numerical integration or analogue techniques.

Let us consider the simple arrangement of two rectangular and adjoining areas at right angles as shown in Fig. 6.5 and determine the shape factor F_{12} using the above formula.

If two incremental areas dA_1 and dA_2 are selected at random on surfaces 1 and 2 the parameters of the above formula become

$$\cos \phi_1 = \frac{z}{r} \; ; \cos \phi_2 = \frac{Y - y}{r}$$

$$r^2 = (x_1 - x_2)^2 + (Y - y)^2 + z^2$$

and substitution yields

$$F_{12} = \frac{1}{XY\pi} \int_0^Z \int_0^Y \int_0^X \int_0^X \frac{z(Y - y)}{[(x_1 - x_2)^2 + (Y - y)^2 + z^2]^2} \, dx_1 \, dx_2 \, dy \, dz$$

Integration with respect to x_1 and x_2 yields

$$F_{12} = \frac{1}{XY\pi} \int_0^Z \int_0^Y \frac{X(Y - y)}{[(Y - y)^2 + z^2]^{3/2}} \tan^{-1} \frac{X}{[(Y - y)^2 + z^2]} \, dy \, dz$$

and with respect to y and z yields

$$\begin{aligned}
F_{12} = \frac{1}{4XY\pi} \Big[& (Y^2 - X^2 + Z^2)\ln(Y^2 + X^2 + Z^2) - (Y^2 + Z^2)\ln(Y^2 + Z^2) \\
& - (Y^2 - X^2)\ln(Y^2 + X^2) + (X^2 - Z^2)\ln(X^2 + Z^2) \\
& + Y^2 \ln Y^2 + Z^2 \ln Z^2 - X^2 \ln X^2 + 4XZ \tan^{-1}\left(\frac{X}{Z}\right) \\
& + 4YX \tan^{-1}\left(\frac{X}{Y}\right) - 4X(Y^2 + Z^2)^{1/2} \tan^{-1}\left(\frac{X}{(Y^2 + Z^2)^{1/2}}\right) \Big]
\end{aligned}$$

$$(6.4)$$

Equations of this type are rather cumbersome for practical use and are therefore often represented in graphical form. Equation (6.4) is presented graphically in Fig. 6.6 and Fig. 6.7 gives the shape factor for parallel rec-

An Introduction to the Analysis of Heat Flow

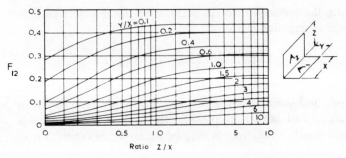

Figure 6.6 Shape factor chart for adjacent rectangles.

tangles. Shape factor charts are available for other configurations in radiation texts such as Hottel and Sarofim (1967) and Sparrow and Cess (1966). For the case of radiation between adjacent squares at right angles, $X = Y = Z$ and also the sides may be equated to unity as physically similar arrangements have the same shape factor. Substitution into equation (6.4) then gives

$$F_{12} = \frac{1}{4\pi}\left[\ln 3 - 2\ln 2 + 8\tan^{-1}(1) - 4\sqrt{2}\tan^{-1}\left(\frac{1}{\sqrt{2}}\right)\right] = 0.2 \text{ (exactly)}$$

It follows that for radiation within a hollow cube with black internal surfaces the heat transfer from one side is equally divided among the other five sides.

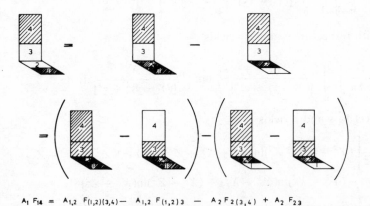

$$A_1 F_{14} = A_{1,2} F_{(1,2)(3,4)} - A_{1,2} F_{(1,2)3} - A_2 F_{2(3,4)} + A_2 F_{23}$$

Figure 6.7 Shape factor chart for parallel rectangles.

In some situations a knowledge of the shape factor for one configuration can be used to calculate the shape factor for a more complex configuration. Consider the case of radiation from surface 1 to 3 for the arrangement shown in Fig. 6.8. Although charts of the shape factor for this arrangement may not be available it can be easily ascertained by subtracting the shape factor for radiation from 1 to 2 from the shape factor for radiation from 1 to 2 and 3

98

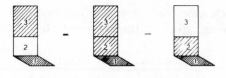

$$F_{13} \quad - \quad F_{1(2,3)} \quad - \quad F_{12}$$

Figure 6.8 Shape factor manipulation.

combined, as shown diagrammatically in the figure. The shape factors for surfaces 1 to 2 and 1 to 2, 3 can be found from the data of Fig. 6.6. The solution of a slightly more complex arrangement is outlined in Fig. 6.9

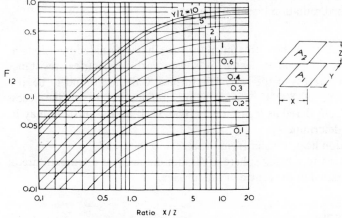

Ratio X/Z

Figure 6.9 Shape factor manipulation.

where the shape factor between areas 1 and 4 is required. In this case radiation from areas 1 and 2 (rather than area 1 only as in the previous case) is involved and the area of the surface must therefore be included in the manipulation. Similar manipulation is possible for rectangles which are perpendicular to each other but not in line as indicated in Fig. 6.10. By considering incremental areas in opposing rectangles and deriving an equation similar to that obtained for adjacent rectangles it can be shown (see Welty, 1974) that

$$A_1 F_{14} = A_2 F_{23} = A_3 F_{32} = A_4 F_{41} \tag{6.5}$$

In principle it is possible to find the radiation shape factor between any two rectangular areas in a perpendicular or parallel plane to each other at any

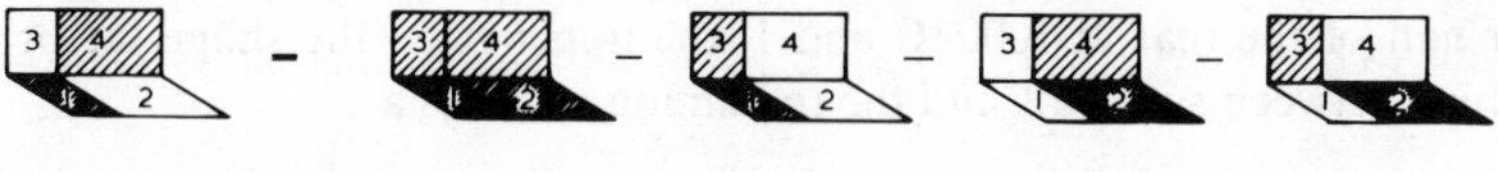

$$A_1 F_{14} \quad = \quad A_{1,2} F_{(1,2)(3,4)} \quad - \quad A_1 F_{13} \quad - \quad A_2 F_{24} \quad - \quad A_2 F_{23}$$

But $A_2 F_{23} = A_1 F_{14}$ (equation 6.5)

$$\therefore A_1 F_{14} \quad = \quad \tfrac{1}{2}(A_{1,2} F_{(1,2)(3,4)} - A_1 F_{13} - A_2 F_{24})$$

Figure 6.10 Shape factor manipulation.

known distance from the axis. In some cases the calculation involved can become tedious and generalized forms are given by Hamilton and Morgan (1952).

For more complicated configurations the shape factor may be determined by empirical or semi-empirical methods. One method involves projection of the shape of the area on to the inner shell of a translucent hemisphere. The hemisphere with the shadow of the area on it is then photographed at a suitable angle and the shadow area determined (see Eckert and Drake, 1972). Semi-empirical approaches include the crossed-string method and the Monte Carlo statistical method, further details of which are available in the previously mentioned radiation heat-transfer texts.

Example 6.2. A large heat treatment furnace has internal dimensions as indicated in Fig. 6.11 and is heated by silicon carbide electrical resistance heaters fitted in a recess at the rear. The furnace walls are found to be at a mean temperature of 800 °C when the mean temperature of the recess (which may be considered as a surface) is 1000 °C. If all the surfaces may be assumed black determine:

   (a) the radiation heat transfer from the recess to each wall,
   (b) the radiation to a black sphere of 40 mm diameter at a temperature of 800 °C situated in the centre of the furnace.

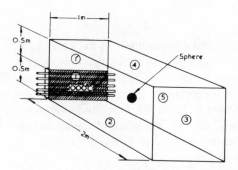

Figure 6.11   Internal dimensions of the furnace.

Solution. (a) With the surfaces numbered as in Fig. 6.11 (with 1 referring to the lower half of the rear at 1000 °C and 1' the upper part) the shape factor relationships between surface 1 and the remaining surfaces are:

F_{12}: available from Fig. 6.6.

F_{13}: $\qquad\qquad A_1 F_{13} = A_3 F_{31}$

also $\qquad\qquad A_1 F_{13} = A_1 F_{1'3}$ by symmetry

and $\qquad\qquad A_1 F_{1'3} = A_3 F_{31'}$

By summation and substitution

$$A_3 F_{3(1,1')} = A_3 F_{31} + A_3 F_{31'}$$
$$= A_1 F_{13} + A_{1'} F_{1'3}$$
$$= 2 A_1 F_{13}$$

therefore
$$F_{13} = \frac{1}{2} \frac{A_3}{A_1} F_{3(1,1')}$$

where $F_{3(1,1')}$ is the shape factor of the total rear surface from the front surface and is available from Fig. 6.7.

F_{14}:
$$A_1 F_{14} = A_{1,1'} F_{(1,1')4} - A_{1'} F_{1'4}$$

and both shape factors on the right are available from Fig. 6.6.

F_{15}:
$$A_1 F_{15} = A_{1,1'} F_{(1,1')5} - A_{1'} F_{1'5}$$

and by symmetry
$$A_1 F_{15} = A_{1'} F_{1'5}$$

therefore
$$A_1 F_{15} = \tfrac{1}{2} A_{1,1'} F_{(1,1')5}$$

where $F_{(1,1')5}$ is available from Fig. 6.6.

Radiation to the wall opposite 5 is the same as to surface 5. Substitution of shape factors from the charts and the appropriate areas yields:

$$F_{12} = 0.32$$
$$F_{13} = 0.07$$
$$F_{14} = 0.14$$
$$F_{15} = 0.23$$

The heat transfer is given in general form by $Q_{mn} = -A_m F_{mn}(T_2^4 - T_1^4)\sigma$ for each black surface and substitution of $T_2 = 1073 °K$, $T_1 = 1273 °K$ and $\sigma = 56.7 \times 10^{-12}$ kW/m^2 K^4 yields the required heat transfer rates:

$$Q_{12} = 11.8 \text{ kW}$$
$$Q_{13} = 2.58 \text{ kW}$$
$$Q_{14} = 5.16 \text{ kW}$$
$$Q_{15} = 8.48 \text{ kW}$$

(Since area 1 is completely enclosed by the other surfaces it follows that

$$F_{12} + F_{13} + F_{14} + 2F_{15} = 1$$

Substitution for the shape factors yields a sum of 0.99 and the 1 % error is due to inaccuracies in reading the charts.)

An Introduction to the Analysis of Heat Flow

(b) Calculation of the exact shape factor of the sphere from surface 1 would involve a lengthy integration. Some estimation of the shape factor may be made by considering radiation radially from the sphere to the surroundings. Since radiation from the sphere is the same in each direction the shape factor may be approximately equated to the ratio of the projected area of surface 1 to the total area:

$$F_{s1} \approx \frac{A_1 \cos \phi}{4\pi x^2}$$

where ϕ is the angle between a line from the sphere to the centre of surface 1 and the normal to the surface and x is the length of this line. There is no heat transfer from the sphere to the rest of the surroundings as the temperature is uniform and therefore:

$$Q_{1s} = -A_s F_{s1} \sigma (T_s^4 - T_1^4)$$

$$= -4\pi(0.02)^2 \times \frac{0.5 \times 0.971}{4\pi(1.03)^2} \times 56.7 \times \left[\left(\frac{1073}{1000}\right)^4 - \left(\frac{1273}{1000}\right)^4 \right]$$

$$= 0.0134 \text{ kW}$$

Thus 13.4 watts are transferred by radiation alone from surface 1 to the sphere under the non-equilibrium conditions specified.

6.4 Examples involving Radiation Networks

In many practical problems neither the simplification of parallel surfaces ($F = 1$) nor the simplification of black surfaces ($\varepsilon = 1$) is appropriate and a more general analysis is required. A popular method is to make use of an electrical analogy consisting of a network of resistances to represent the surface and shape restrictions to heat flow. Expressions for these resistances were derived in Section 1.4 and may be summarized as follows.

Surface resistance of surface 1: $\dfrac{1 - \varepsilon_1}{\varepsilon_1 A_1}$

Shape resistance of surface 2 as seen from surface 1: $\dfrac{1}{A_1 F_{12}}$

The application of this method to two parallel grey surfaces was demonstrated in Section 6.2. Its application to more complex arrangements is illustrated by the following examples.

Example 6.3. Two parallel discs of 1 m diameter are situated 2 m apart in surroundings at a room temperature of 20 °C. The inner side of one disc has an emissivity of 0.5 and is maintained at 500 °C by electrical resistance heating, and the outer side of the disc is well insulated. The other disc is open to radiation both sides and reaches an equilibrium temperature. Calculate this equilibrium temperature and the heat flow rate from the first disc (assuming heat transfer is entirely by radiation). Discuss the effect on the solution if both sides of the second disc are perfect mirrors.

Solution. There are effectively three surfaces involved in the problem; the hot disc at 500 °C, the cool disc at an unknown temperature T_2 and the surroundings at 20 °C. The general network for three surfaces that 'see' each other is shown in Fig. 6.12c. Note that each surface has a surface resistance

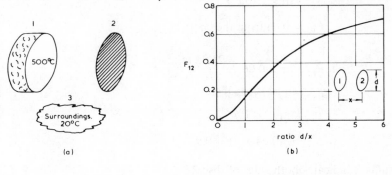

(a)   (b)

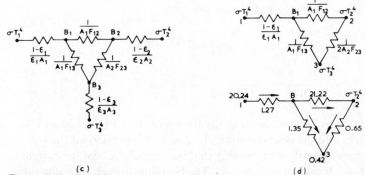

(c)   (d)

Figure 6.12   Radiation between parallel discs.

and that the shape resistances are arranged so that each surface is connected to the other two. In this particular case some simplifications to the network can be made. Surroundings are generally assumed to be effectively black ('effectively' because distant non-black surroundings act as black surfaces as no radiation is reflected) and ε_3 is therefore unity. The surface resistance of 3 is therefore zero and $B_3 = \sigma T_3^4$. The emissivity of surface 2 is not given but it is known that the disc is in thermal equilibrium and that heat exchange with it is entirely by radiation. Since there is no *net* heat transfer from the disc there is no heat (or current) flow in the leg from B_2 to σT_2^4, or alternatively the heat flow from 1 to 2 is equal to the heat flow from 2 to 3. Under this condition the potential at B_2 is the same as that at σT_2^4 and the circuit may be simplified

103

to that shown in Fig. 6.12d. The parameters involved may be found as follows:

$$\sigma T_1^4 = 56.7 \times \left(\frac{773}{1000}\right)^4 = 20.24 \text{ kW/m}^2$$

$$\sigma T_3^4 = 56.7 \times \left(\frac{293}{1000}\right)^4 = 0.418 \text{ kW/m}^2$$

$$\frac{1 - \varepsilon_1}{\varepsilon_1 A_1} = \frac{1 - 0.5}{0.5 \times \pi/4 \times 1} = 1.27 \text{ m}^{-2}$$

$$F_{12} = 0.06 \text{ from the chart.}$$

$$\frac{1}{A_1 F_{12}} = \frac{4}{\pi \times 0.06} = 21.22 \text{ m}^{-2}$$

$$F_{13} = 1 - F_{12} = 0.94$$

$$F_{23} = \tfrac{1}{2}(F_{2'3} + F_{2''3})$$

$$= \tfrac{1}{2}(0.94 + 1) = 0.97$$

where: suffix 2 indicates both sides of disc 2

suffix 2′ indicates the left-hand side only

suffix 2″ indicates the right-hand side only

$$\frac{1}{A_1 F_{13}} = \frac{4}{\pi \times 0.94} = 1.35 \text{ m}^{-2}$$

$$\frac{1}{A_2 F_{23}} = \frac{4}{2\pi \times 0.97} = 0.65 \text{ m}^{-2}$$

The circuit may now be considered in the form shown in Fig. 6.12d and the problem resolves into the determination of B_1 and σT_2^4 from circuit theory.

The circuit may be analysed by noting that the sum of the heat (or current) flows entering and leaving a node is zero. The heat flow in each leg is equal to the potential drop divided by the resistance.

At point B:

$$Q_{1-B} + Q_{2-B} + Q_{3-B} = 0$$

$$\frac{20.24 - B}{1.27} + \frac{\sigma T_2^4 - B}{21.22} + \frac{0.418 - B}{1.35} = 0$$

At point 2:

$$\frac{B - \sigma T_2^4}{21.22} + \frac{0.418 - \sigma T_2^4}{0.65} = 0$$

Solution of these simultaneous equations yields

$$\sigma T_2^4 = 0.715 \text{ kW/m}^2$$

$$B = 10.34 \text{ kW/m}^2$$

from which

$$T_2 = 335.1 \,°K = \underline{62\,°C}$$

and the heat flow from the hot disc is given by:

$$Q_{1-B} = \frac{\sigma T_1^4 - B}{1.27}$$

$$= \frac{20.24 - 10.34}{1.27}$$

$$\underline{Q_{1-B} = 7.79 \text{ kW}}$$

As the heat energy leaving disc 1 is all directly or indirectly passed to the surroundings we may compare the above value with the heat received by the surroundings:

$$Q_{1-B} = Q_{B-3} + Q_{2-3}$$

$$= \frac{10.34 - 0.42}{1.35} + \frac{0.715 - 0.42}{0.65}$$

$$= 7.79 \text{ kW}$$

The second disc therefore has no effect on heat flow from the hot disc and simply reradiates the heat energy which falls on it to the surroundings. This applies at any value of emissivity of the second disc and in particular when $\varepsilon = 1$ and when $\varepsilon = 0$ (both sides). In the former case the disc is black and not only absorbs all the radiation that falls on it but also emits all this radiation if it is in thermal equilibrium. In the latter case the disc is a perfect mirror and reflects all the radiation that falls on it. The solution to the problem is therefore unchanged if the disc is a perfect mirror each side because, whatever the surface, all the incident energy is reradiated when in thermal equilibrium. By similar reasoning applied to the more practical example of the domestic coal fire it can be shown that there would be no overall advantage to replacing the firebricks at the back of the hearth by polished metal sheets.

Example 6.4. The entrance to a large departmental store is fitted with an internal canopy. The ceiling of the canopy is maintained at 107 °C by a heating unit and the surface emissivity is 0.9. The physical dimensions of the canopy are as indicated in Fig. 6.13 and the sides and floor may be considered as thermal insulators at uniform temperatures.

Part of an analysis of heat losses from the store involves a determination of the radiation heat transfer through the open entrance. It is required to

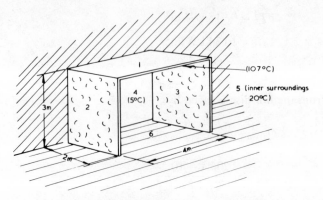

Figure 6.13 The door canopy.

estimate this radiation under typical winter conditions when the outside temperature is 5 °C and the store is at 20 °C.

Solution. The radiation network for the general case where six surfaces, each with a surface resistance, 'see' each other is shown in Fig. 6.14. The central group of links are all space resistances of the form $1/A_m F_{mn}$ while the six projecting arms are surface resistances of the form $(1 - \varepsilon_m)/\varepsilon_m A_m$. The circuit may be simplified in this case by noting the following points (where sides are numbered as in Fig. 6.13):

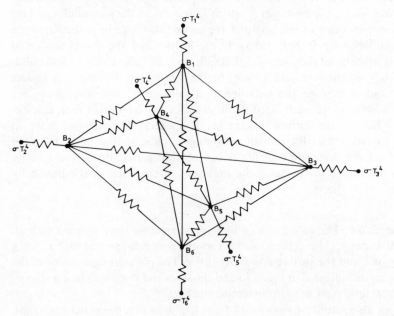

Figure 6.14 Radiation network for six surfaces.

106

(i) Owing to symmetry $T_2 = T_3$ and there is no radiation (and therefore no circuit link) between 2 and 3.

(ii) If the surroundings both outside and inside the store are assumed to be effectively black there are no surface resistances for sides 4 and 5.

(iii) Thermally insulated surfaces reradiate all the incident radiation falling on them when in thermal equilibrium as discussed earlier. The heat flow in arms 2, 3 and 6 is therefore zero.

The resistances of the remaining elements of the circuit are as follows, (where shape factors for each configuration are obtained from the charts of Figs. 6.6 and 6.7).
Shape resistances (in units of m^{-2}):

(a)
$$\frac{1}{A_1 F_{12}} = \frac{1}{A_1 F_{13}} = \frac{1}{8 \times 0.135} = 0.926$$

$$\frac{1}{A_1 F_{14}} = \frac{1}{A_1 F_{15}} = \frac{1}{8 \times 0.27} = 0.463$$

$$\frac{1}{A_1 F_{16}} = \frac{1}{8 \times 0.18} = 0.694$$

(b)
$$\frac{1}{A_2 F_{21}} = \frac{1}{A_2 F_{26}}$$

$$= \frac{1}{A_1 F_{12}} = 0.926$$

$$\frac{1}{A_2 F_{23}} \quad \text{not required}$$

$$\frac{1}{A_2 F_{24}} = \frac{1}{A_2 F_{25}} = \frac{1}{6 \times 0.27} = 0.617$$

(c)
$$\frac{1}{A_4 F_{41}} = \frac{1}{A_4 F_{46}}$$

$$= \frac{1}{A_1 F_{14}} = 0.463$$

$$\frac{1}{A_4 F_{42}} = \frac{1}{A_4 F_{43}}$$

$$= \frac{1}{A_2 F_{24}} = 0.617$$

$$\frac{1}{A_4 F_{45}} = \frac{1}{12 \times 0.36} = 0.231$$

107

$$\frac{1}{A_3 F_{3n}} \quad \text{similar to (b)}$$

$$\frac{1}{A_5 F_{5n}} \quad \text{similar to (c)}$$

$$\frac{1}{A_6 F_{6n}} \quad \text{similar to (a)}$$

Surface resistances:

$$\frac{1-\varepsilon_1}{\varepsilon_1 A_1} = \frac{1-0.9}{0.9 \times 8} = 0.0139 \text{ m}^{-2}$$

all other surface resistances are accounted for by (ii) and (iii).
Thermal potentials:

$$\sigma T_1^4 = 56.7 \left(\frac{380}{1000}\right)^4 = 1.185 \text{ kW/m}^2$$

$$\sigma T_2^4 = \sigma T_3^4 = \sigma T_s^4$$

$$\sigma T_4^4 = 56.7 \left(\frac{278}{1000}\right)^4 = 0.387 \text{ kW/m}^2$$

$$\sigma T_5^4 = 56.7 \left(\frac{293}{1000}\right)^4 = 0.418 \text{ kW/m}^2$$

The circuit is now as shown in Fig. 6.15. The unknown node potentials are B_1, σT_s^4 and σT_6^4 and these may be found by summing the heat flow at each node, (where the heat flow is the potential difference divided by the resistance of each leg).

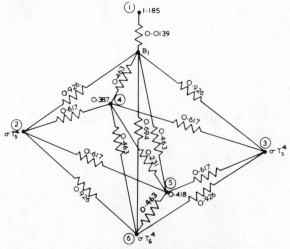

Figure 6.15 Radiation network for Example 6.4.

At node B_1:

$$\frac{1.185 - B_1}{0.0139} + 2\left(\frac{\sigma T_s^4 - B_1}{0.926}\right) + \frac{0.387 - B_1}{0.463} + \frac{0.418 - B_1}{0.463} + \frac{\sigma T_6^4 - B_1}{0.694} = 0$$

At node 2:

$$\frac{B_1 - \sigma T_s^4}{0.926} + \frac{0.387 - \sigma T_s^4}{0.617} + \frac{0.418 - \sigma T_s^4}{0.617} + \frac{\sigma T_6^4 - \sigma T_s^4}{0.926} = 0$$

At node 6:

$$\frac{B_1 - \sigma T_6^4}{0.694} + 2\left(\frac{\sigma T_s^4 - \sigma T_6^4}{0.926}\right) + \frac{0.387 - \sigma T_6^4}{0.463} + \frac{0.418 - \sigma T_6^4}{0.463} = 0$$

Solution of these three simultaneous equations yields

$$B_1 = 1.115 \text{ kW/m}^2$$
$$\sigma T_s^4 = 0.580 \text{ kW/m}^2$$
$$\sigma T_6^4 = 0.580 \text{ kW/m}^2$$

from which the temperature of the insulated surfaces is 318 °K or 45 °C. The required radiation heat flow through the door (to 'surface' 4) is given by

$$Q_4 = \frac{B_1 - 0.387}{0.463} + 2\left(\frac{\sigma T_s^4 - 0.387}{0.617}\right) + \frac{0.418 - 0.387}{0.231} + \frac{\sigma T_6^4 - 0.387}{0.463}$$

$$= 1.572 + 0.626 + 0.134 + 0.417$$

$$\underline{Q_4 = 2.75 \text{ kW}}$$

The answer may be checked by calculating Q_5 and Q_1 and substituting in $Q_1 = Q_4 + Q_5$. By similar calculation to the above Q_5, the net heat flow into the store, is found to be 2.25 kW.

$$Q_1 = \frac{1.185 - B_1}{0.0139} = 5.03 \text{ kW}$$

Therefore:

$$Q_4 = Q_1 - Q_5$$

$$\underline{2.78 \text{ kW}}$$

In more complex cases matrix methods and computation using a digital computer are obviously desirable. In the situation studied in this example radiation would be a minor heat transfer mechanism compared to the heat transport caused by the exchange of cold and warm air at the door.

Example 6.5 (involving gas radiation). The temperature of the exhaust gases of an internal combustion engine is 400 °C and (from measurements of the proportion of water vapour and carbon dioxide in the gases and with the aid of appropriate gas radiation tables) the emissivity of the gases is estimated as 0.2 at this temperature. In order to decrease pollution from the gases they are passed through a pollution controller part of which consists of a thin annulus of 100 mm length as shown in Fig. 6.16. Both surfaces of the annulus

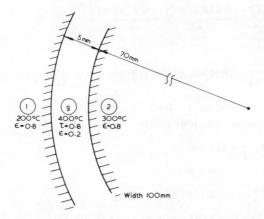

Figure 6.16   Absorbing gas enclosed in an annulus.

have an emissivity of 0.8 and the temperatures of the inner and outer surfaces are 300 °C and 200 °C. The flow rate of the gases is high and therefore their temperature does not drop substantially through the annulus.

Determine the total heat flow rate by radiation to the outer surface and compare this value with that obtained if the gas emissivity is ignored. In addition, find the heat flow rate by radiation from the gas.

Solution. Although it is beyond the scope of this text to consider gas radiation in any detail, simple problems involving gas radiation may be solved using the network analogy. The reflectivity of gases can generally be taken as zero and therefore $\alpha + \tau = 1$. In this case the emissivity and therefore the absorptivity is 0.2 and so the transmissivity is 0.8. That is to say 0.8 of the direct radiation between surfaces 2 and 1 which would be transferred without the gas present is actually transferred, or, expressed another way, the shape resistance is increased by a factor of $1/0.8$. The shape resistance through the gas is therefore $1/(A_2 F_{21} \tau_g)$. In addition there is radiation between the gas itself and the surfaces. The proportion of radiation to the gas from surface 2 which is absorbed by the gas is 0.2 and the shape resistance to the gas is therefore increased by a factor of $1/0.2$. That is to say the shape resistance to the gas may be expressed as $1/(A_2 F_{2g} \varepsilon_g)$. The surface resistance of the gas is obviously zero and the network analogy for the system is as shown in Fig. 6.17.

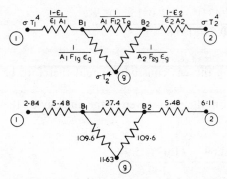

Figure 6.17 Radiation network for problem 6.5.

The network resistance may be calculated as follows:

$$\frac{1 - \varepsilon_1}{\varepsilon_1 A_1} \approx \frac{1 - \varepsilon_2}{\varepsilon_2 A_2} \approx \frac{1 - 0.8}{0.8 \times 0.0456} = 5.48 \text{ m}^{-2}$$

(where the surface may be considered as two parallel plates of area $2\pi \times 0.0725 \times 0.1 = 0.0456 \text{ m}^2$, as the gap is small compared to the diameter).

$$\frac{1}{A_1 F_{12} \tau_g} = \frac{1}{0.0456 \times 1 \times 0.8} = 27.4 \text{ m}^{-2}$$

$$\frac{1}{A_1 F_{1g} \varepsilon_g} \approx \frac{1}{A_2 F_{2g} \varepsilon_g} = \frac{1}{0.0456 \times 1 \times 0.2} = 109.6 \text{ m}^{-2}$$

The potentials are:

$$\sigma T_1^4 = 56.7 \left(\frac{473}{1000} \right)^4 = 2.84 \text{ kW/m}^2$$

$$\sigma T_2^4 = 56.7 \left(\frac{573}{1000} \right)^4 = 6.11 \text{ kW/m}^2$$

$$\sigma T_g^4 = 56.7 \left(\frac{673}{1000} \right)^4 = 11.63 \text{ kW/m}^2$$

Solution of the network as in previous problems yields

$$B_1 = 3.69 \text{ kW/m}^2$$

$$B_2 = 5.94 \text{ kW/m}^2$$

and the required heat flow rate to the outer surface is

$$Q_1 = \frac{3.69 - 2.84}{5.48} = 0.155 \text{ kW}$$

The heat flow rate to the outer surface by radiation, ignoring the gas, may be found from equation (6.3):

$$(Q_1)_{\text{no gas}} = -\frac{56.7 \times 0.0456[(473)^4 - (573)^4] \times 10^{-12}}{1/0.8 + 1/0.8 - 1}$$

$$= 0.099 \text{ kW}$$

and the error involved by neglecting the gas emissivity would therefore be considerable.

The heat flow rate by radiation from the gas is given in the analogy by the heat flow from point g to B_1 and B_2:

$$Q_g = \frac{11.63 - 3.69}{109.6} + \frac{11.63 - 5.94}{109.6}$$

$$Q_g = 0.124 \text{ kW}$$

These answers may be checked by noting that the heat flow to the outer surface is equal to the sum of the heat flows from the gas and the inner surface.

Conduction Analysis 7

7.1   Introduction

The analysis of heat flow by conduction through a material involves determination of the temperature distribution. In Chapter 2 some cases involving a simple temperature distribution were considered. In general terms the distribution is described by a partial differential equation and theoretical analysis of the heat flow depends upon the possibility of solution of this equation. Once the temperature distribution and boundary conditions have been specified, çonduction problems therefore resolve into the mathematical problems of differential equation analysis involving separation of variables, Laplace transformations and Bessel functions. In this chapter the setting-up of the differential equation for the temperature distribution will be described but the more advanced mathematical techniques of solution will not be covered. These techniques and the solutions for many common conduction problems are adequately detailed in texts such as Carslaw and Jaeger (1959), Schneider (1955) and Myers (1971). Many practical problems do not lend themselves to an analytical approach and numerical analysis using digital computers, electrical analogues or constructional techniques are employed.

In Chapter 2 the Fourier equation for one-dimensional conduction through a plane wall was introduced as

$$Q = -kA \frac{\mathrm{d}T}{\mathrm{d}x} \tag{7.1}$$

The general conduction equation in Cartesian coordinates (including the effects of heat generation within the material and temperature change with time) may be determined by applying equation (7.1) in each Cartesian direction through the cube, as shown in Fig. 7.1, and performing an energy

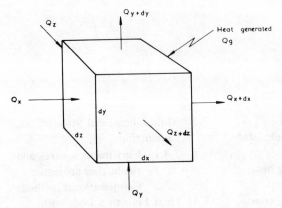

Figure 7.1 Conduction through a cubic element.

balance. Equating the difference between the heat flow rates into and out of the cube to the rate of change of internal energy U_i with time t yields

$$(Q_x + Q_y + Q_z + Q_g) - (Q_{x+dx} + Q_{y+dy} + Q_{z+dz}) = \frac{\mathrm{d}U_i}{\mathrm{d}t}$$

The terms of this equation may be expressed as follows:

$$Q_x = -k\,\mathrm{d}y\,\mathrm{d}z\,\frac{\partial T}{\partial x}$$

$$Q_{x+dx} = -\left[k\,\frac{\partial T}{\partial x} + \frac{\partial}{\partial x}\left(k\,\frac{\partial T}{\partial x}\right)\mathrm{d}x\right]\mathrm{d}y\,\mathrm{d}z$$

and similarly for Q_y, Q_{y+dy}, Q_z and Q_{z+dz}

$$Q_g = q_g\,\mathrm{d}x\,\mathrm{d}y\,\mathrm{d}z$$

where q_g is the heat energy generated per unit volume.

$$\frac{\mathrm{d}U_i}{\mathrm{d}t} = \rho c\,\mathrm{d}x\,\mathrm{d}y\,\mathrm{d}z\,\frac{\partial T}{\partial t}$$

where ρ and c are the density and specific heat.
Substitution then gives

$$\frac{\partial}{\partial x}\left(k\,\frac{\partial T}{\partial x}\right) + \frac{\partial}{\partial y}\left(k\,\frac{\partial T}{\partial y}\right) + \frac{\partial}{\partial z}\left(k\,\frac{\partial T}{\partial z}\right) + q_g = \rho c\,\frac{\partial T}{\partial t} \qquad (7.2)$$

and for constant k throughout the material this becomes

$$\frac{\partial^2 T}{\partial x^2} + \frac{\partial^2 T}{\partial y^2} + \frac{\partial^2 T}{\partial z^2} + \frac{q_g}{k} = \frac{1}{\alpha}\frac{\partial T}{\partial t} \qquad (7.3)$$

where α is the thermal diffusivity defined as $\alpha = k/\rho c$ (with units of m^2/s). The thermal diffusivity indicates the rate at which heat is distributed in a material and this rate depends not only on the conductivity but also on the rate at which the energy can be stored, as is evident from the product ρc in the denominator.

In many situations equation (7.3) may be simplified because there is no dependence of the temperature on time. In this case conduction is said to occur in the 'steady-state' and the heat flow equation reduces to Poisson's equation:

$$\frac{\partial^2 T}{\partial x^2} + \frac{\partial^2 T}{\partial y^2} + \frac{\partial^2 T}{\partial z^2} + \frac{q_g}{k} = 0 \qquad (7.4)$$

In the absence of internal heat generation it further reduces to Laplace's equation:

$$\frac{\partial^2 T}{\partial x^2} + \frac{\partial^2 T}{\partial y^2} + \frac{\partial^2 T}{\partial z^2} = 0 \qquad (7.5)$$

The heat conduction equation in radial coordinates may be found in a similar way by an energy balance on an annular ring as shown in Fig. 7.2.

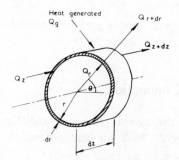

Figure 7.2 Conduction through an annular element.

For simplicity it is assumed that there is complete symmetry about the z axis. An energy balance then yields

$$Q_r + Q_z - (Q_{r+dr} + Q_{z+dz}) + Q_g = \frac{dU_i}{\partial t}$$

and the terms are given by

$$Q_r = -2\pi r\, dz k \frac{\partial T}{\partial r}$$

$$Q_z = -2\pi r\, dr k \frac{\partial T}{\partial z}$$

115

$$Q_{r+dr} = -2\pi(r + dr)\, dz \left[k\frac{\partial T}{\partial r} + \frac{\partial}{\partial r}\left(k\frac{\partial T}{\partial r} \right) dr \right]$$

$$Q_{z+dz} = -2\pi r\, dr \left[k\frac{\partial T}{\partial z} + \frac{\partial}{\partial z}\left(k\frac{\partial T}{\partial z} \right) dz \right]$$

$$Q_g = q_g 2\pi r\, dr\, dz$$

$$\frac{\partial U_i}{\partial t} = \rho c 2\pi r\, dr\, dz\, \frac{\partial T}{\partial t}$$

Substitution, neglecting second-order differentials, yields

$$r\frac{\partial}{\partial r}\left(k\frac{\partial T}{\partial r} \right) + k\frac{\partial T}{\partial r} + r\frac{\partial}{\partial z}\left(k\frac{\partial T}{\partial z} \right) + rq_g = \rho c r\frac{\partial T}{\partial t}$$

and for constant thermal conductivity

$$\frac{\partial^2 T}{\partial r^2} + \frac{1}{r}\frac{\partial T}{\partial r} + \frac{\partial^2 T}{\partial z^2} + \frac{q_g}{k} = \frac{1}{\alpha}\frac{\partial T}{\partial t} \tag{7.6}$$

For the steady-state case with no heat generation or variation in the z direction this reduces to

$$\frac{\partial^2 T}{\partial r^2} + \frac{1}{r}\frac{\partial T}{\partial r} = 0 \tag{7.7}$$

In the following sections we shall consider some applications and methods of solution of these equations.

7.2 One-Dimensional Steady-State Conduction

Steady-state conduction in one dimension through a plate and radially through a cylinder were studied in Chapter 2. With no heat generation equation (7.5) reduces to

$$\frac{\partial^2 T}{\partial x^2} = 0 \tag{7.8}$$

giving

$$T = Mx + N$$

where M and N are constants.

The temperature distribution is therefore linear and the heat flux is constant throughout the plate. For the radial case equation (7.7) yields, (by introducing the dummy variable $v = dT/dr$)

$$T = M \ln r + N$$

and the temperature distribution is therefore logarithmic in form. Some cases which are more complicated will now be considered.

7.2.1 Cases involving heat generation

Heat generation (by radioactive decay or the passage of electric current for example) leads to a one-dimensional temperature distribution which is non-linear in form. Consider heat flow along a bar between two thermal reservoirs at temperatures T_1 and T_2 as shown in Fig. 7.3. The bar is perfectly insulated

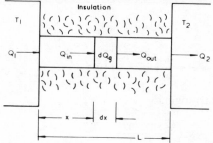

Figure 7.3 Conduction through a bar with heat generation.

and heat is generated uniformly within it at the rate of q_g *per unit volume*. Heat flow into the element is given by

$$Q_{in} = -kA \frac{dT}{dx}$$

and out of the element by

$$Q_{out} = -kA \left[\frac{dT}{dx} + \frac{d^2T}{dx^2} dx \right]$$

where k and A are the conductivity and cross-sectional area of the bar. Heat generated within the element is

$$dQ_g = q_g A\, dx$$

and substitution in the heat balance

$$Q_{out} - Q_{in} = dQ_g$$

gives

$$\frac{d^2T}{dx^2} = -\frac{q_g}{k} \tag{7.9}$$

which is equation (7.4) reduced to one-directional heat flow. Double integration gives

$$T = \frac{-q_g}{k} \frac{x^2}{2} + Mx + N$$

and substitution of the boundary conditions

$$T = T_1 \text{ at } x = 0$$

$$T = T_2 \text{ at } x = L$$

gives

$$T = T_1 + \frac{(T_2 - T_1)x}{L} + \frac{q_g}{k}\frac{(L - x)x}{2} \qquad (7.10)$$

The proportion of total heat, generated within the bar, which flows out at each end may be found as follows. The heat flow along the bar is given by

$$Q = -kA\frac{dT}{dx}$$

$$= -kA\left[\frac{T_2 - T_1}{L} + \frac{q_g}{k}\left(\frac{L}{2} - x\right)\right]$$

Substituting $Q = Q_1$ at $x = 0$ and rearranging:

$$Q_1 + \frac{q_g AL}{2} = -kA\frac{(T_2 - T_1)}{L}$$

and since $q_g AL$ is the total heat Q_g generated in the bar

$$Q_1 + \frac{Q_g}{2} = -kA\frac{(T_2 - T_1)}{L} \qquad (7.11)$$

The right-hand side of this equation is equal to the heat flux along the bar under conditions when no heat is generated and in this case therefore the effect of the generated heat is to decrease the heat flow into the bar at side 1 by $Q_g/2$ and to increase the heat flow out of the bar at side 2 by $Q_g/2$. In other words, for the situation shown in Fig. 7.3, half the generated heat flows out at each end of the bar whatever the values of the reservoir temperatures.

If the heat generated within a material is due to the passage of an electric current, q_g may be expressed in electrical terms by using the relationships:

$$Q_g = I^2 R$$

$$R = \frac{\rho L}{A}$$

and

$$q_g = \frac{Q_g}{AL}$$

where I is the current, R is electrical resistance, ρ is resistivity, and L and A are the length and cross-sectional area of the conductor. Combination of these relationships leads to

$$q_g = \left(\frac{I}{A}\right)^2 \rho = i^2\rho \qquad (7.12)$$

where i is the current density.

Example 7.1. A long stainless steel bar of 20 mm × 20 mm square cross-section is perfectly insulated on three sides and is maintained at a temperature

118

of 400 °C on the remaining side. Determine the maximum temperature in the bar when it is conducting a current of 1000 amps. The thermal and electrical conductivity of stainless steel may be taken as 16 W/m K and 1.5×10^4 (ohm-cm)$^{-1}$ and the heat flow at the ends may be neglected.

Solution. Since heat can only flow out of the bar at the uninsulated side the maximum temperature is at the opposite side. Let x be measured across the bar from the side at maximum temperature towards the uninsulated side. The equation for the temperature distribution is

$$\frac{d^2T}{dx^2} = -\frac{q_g}{k}$$

and double integration yields

$$T = -\frac{q_g}{k}\frac{x^2}{2} + Mx + N$$

From equation (7.12)

$$q_g = \left(\frac{I}{A}\right)^2 \rho$$

and substitution noting that ρ is the reciprocal of electrical conductivity gives

$$q_g = \left(\frac{1000}{4 \times 10^{-4}}\right)^2 \frac{1}{1.5 \times 10^4 \times 10^2 \times 10^3} \frac{kW}{m^3}$$

$$= 4167 \text{ kW/m}^3$$

Substitution of the boundary conditions

$$\text{at} \quad x = 0, \frac{dT}{dx} = 0 \quad (\text{as } q = 0)$$

$$\text{at} \quad x = 0.02 \text{ m}, T = 400 \,°C$$

yields the general equation

$$T = \frac{q_g}{2k}[(0.02)^2 - x^2] + 400$$

Since $T = T_{max}$ when $x = 0$ it follows that

$$T_{max} = \frac{4167 \times (0.02)^2}{2 \times 16 \times 10^{-3}} + 400$$

$$T_{max} = 452.1 \,°C$$

Example 7.2. Derive an equation giving the temperature at the centre of a circular rod conducting electric current in terms of the current density, the wall temperature and the material properties. What is the centre temperature of a stainless steel rod of 20 mm diameter with similar properties to the bar in Example 7.1 and an outer temperature of 400 °C, when conducting 1000 amps?

Solution. For steady-state radial conduction with no variation of the temperature profile along the rod and with a uniformly distributed heat source equation (7.6) may be written in the form

$$r\frac{\partial^2 T}{dr^2} + \frac{dT}{dr} = \frac{-q_g r}{k}$$

and noting that

$$\frac{d}{dr}\left(r\frac{dT}{dr}\right) = r\frac{d^2 T}{dr^2} + \frac{dT}{dr}$$

integration yields

$$r\frac{dT}{dr} = \frac{-q_g r^2}{2k} + M$$

and

$$T = \frac{-q_g r^2}{4k} + M \ln r + N$$

The boundary conditions are: $T = T_s$ at $r = r_s$

$$\frac{dT}{dr} = 0 \qquad \text{at } r = 0 \qquad \text{(from symmetry)}$$

Determination of the constants then yields

$$T = \frac{q_g}{4k}(r_s^2 - r^2) + T_s$$

and if $T = T_0$ at the centre of the rod when $r = 0$

$$T_0 = \frac{q_g r_s^2}{4k} + T_s$$

Substitution of $i^2 \rho$ for q_g then yields the required relationship

$$T_0 = \frac{i^2 \rho r_s^2}{4k} + T_s$$

The centre-temperature of the stainless steel rod may now be found by substitution of the numerical values:

$$T_0 = \left(\frac{1000}{\pi \times 10^{-4}}\right)^2 \frac{10^{-4}}{(1.5 \times 10^4 \times 10^2 \times 10^3)\,4\,(16 \times 10^{-3})} + 400$$

$$= 410.6 \,°C$$

7.2.2 Cases involving convection

Consider heat flow along a bar connecting two thermal reservoirs as discussed in the previous section, but with heat flow by convection from the bar to the surroundings rather than heat generated within the bar. The arrangement is

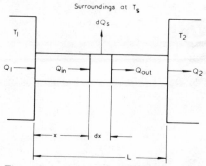

Figure 7.4 Conduction through a bar with heat loss by convection.

shown in Fig. 7.4 and Q_s denotes the heat flow to the surroundings from the perimeter of the bar. The heat flow terms are:

$$Q_{in} = -kA \frac{dT}{dx}$$

$$Q_{out} = -kA \left[\frac{dT}{dx} + \left(\frac{d^2T}{dx} \right) dx \right]$$

and

$$dQ_s = -hP(T_s - T) \, dx$$

where h is the convection heat transfer coefficient, P is the perimeter of the bar and T_s is the temperature of the surroundings. Substitution in the heat balance

$$Q_{out} - Q_{in} = -dQ_s \, .$$

yields

$$-kA \frac{d^2T}{dx^2} = -hP(T - T_s)$$

If $\Delta T = T - T_s$ and $m = \sqrt{\dfrac{hP}{kA}}$ it follows that

$$\frac{d^2 \Delta T}{dx^2} = m^2 \Delta T \qquad (7.13)$$

and the solution to this differential equation is:

$$\Delta T = M \, e^{-mx} + N \, e^{mx} \qquad (7.14)$$

where constants M and N may be determined from the boundary conditions.

An important application of this equation is the cooling fin. In many engineering systems the main resistance to heat flow is the fluid boundary layer and this resistance may be reduced by increasing the surface area using fins. For this reason air-cooled engine cylinders are usually finned and internally or externally finned heat exchanger tubes of the types shown in Fig. 7.5 are available. A good approximation to the temperature distribution

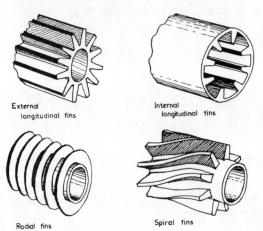

External
longitudinal fins

Internal
longitudinal fins

Radial fins

Spiral fins

Figure 7.5 Tubular fin arrangements.

in a simple fin of rectangular cross-section, long enough for the tip temperature to be close to that of the surroundings, may be obtained by considering an infinitely long fin. The boundary conditions become:

$$\text{at } x = 0, \quad \Delta T = (T_1 - T_s) = \Delta T_1$$

$$\text{at } x = \infty, \ \Delta T = 0$$

and determination of M and N in equation (7.14) yields

$$\Delta T = \Delta T_1 \, e^{-mx}$$

If a fin projecting from a surface has a width w (along the surface) and a thickness t which is very much less than w, the constant m in the above equation becomes

$$m = \sqrt{\frac{hP}{kA}} = \sqrt{\frac{h2(w + t)}{kwt}} \approx \sqrt{\frac{2h}{kt}} \tag{7.15}$$

In practice fins are seldom rectangular and the mathematical solutions for more complex cases are rather involved. The fin's effectiveness in increasing the heat transfer rate is therefore often expressed in terms of a fin efficiency η_f defined as

$$\eta_f = \frac{\text{actual heat transferred}}{\text{heat transferred if the entire fin area is at the temperature of the base of the fin.}}$$

Charts are available which present fin efficiency and fluid flow resistance as functions of the fin parameters, (Jakob, 1949).

Example 7.3. In a chemical process the heat transfer from a surface to distilled water is increased by a number of thin fins, each with dimensions as shown in Fig. 7.6. The metal fins are coated with a 0.1 mm thick layer of

122

plastic to prevent ionization of the water and the ends of the fins are fitted against an insulated wall. The temperature at the base of the fins is 80 °C, the mean water temperature is 20 °C and the heat transfer coefficient between the water and the plastic coating is 0.2 kW/m² K. Determine the temperature at the tip of the fins and the fin efficiency (k for aluminium = 204 W/m K; k for plastic = 0.5 W/m K).

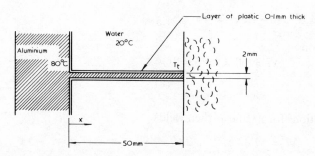

Figure 7.6 The plastic coated fin.

Solution. The temperature distribution in the fin is given by equations (7.13) and (7.14) as

$$\frac{d^2\,\Delta T}{dx^2} = m^2\,\Delta T$$

and

$$\Delta T = M\,e^{-mx} + N\,e^{mx}$$

where m is given by equation (7.15), as

$$m \approx \sqrt{\frac{2h}{kt}}$$

In this case there is the layer of plastic to be taken into account and the total resistance to heat transfer is not only $1/h$ but also the resistance through the plastic, given by t_p/k (where t_p is the coat thickness). The term h in the above expression must therefore be replaced by the overall heat transfer coefficient U given by:

$$\frac{1}{U} = \frac{1}{h} + \frac{t_p}{k}$$

$$= \frac{1}{0.2} + \frac{0.1 \times 10^{-3}}{0.5 \times 10^{-3}} = 5.2$$

$$U = 0.192 \text{ kW/m}^2 \text{ K}$$

Thus

$$m = \sqrt{\frac{2 \times 0.192}{204 \times 10^{-3} \times 2 \times 10^{-3}}} = 30.68 \text{ m}^{-1}$$

123

Substitution of the boundary condition:

$$\frac{dT}{dx} = 0 \text{ at } x = 0.05$$

yields

$$\frac{dT}{dx} = -mM \, e^{-0.05\,m} + mN \, e^{0.05\,m} = 0$$

$$M = N \, e^{0.1\,m} = N \, e^{3.068} = 21.5 \, N$$

The boundary condition

$$\Delta T = 60\,°C \text{ at } x = 0$$

yields

$$60 = M + N$$

Solution of these equations for M and N then yields

$$\Delta T = 57.33 \, e^{-mx} + 2.67 \, e^{mx}$$

and substitution of $m = 30.68$ and $x = 0.05$ at the tip of the fin yields

$$\Delta T = 24.73\,°C$$

The temperature of the tip is therefore $24.73 + 20 = \underline{44.73\,°C}$.

The actual heat transferred to the complete fin per unit width is given by the heat flow Q_b at the base

$$Q_b = -kA \left(\frac{dT}{dx} \right)_{x=0}$$

Now

$$\frac{dT}{dx} = -mM \, e^{-mx} + mN \, e^{mx}$$

and at

$$x = 0 \qquad \frac{dT}{dx} = m(N - M)$$

Thus

$$Q_b = 204 \times 10^{-3} \times (2 \times 10^{-3}) \times 30.68 \times 54.65$$

$$= 0.684 \text{ kW}$$

Alternatively, the total heat transferred from the fin may be found by integrating the heat flow over both sides of the fin surface.

$$Q_b = 2 \int_A U \, \Delta T \, dA$$

$$= 2 \int_0^{x=0.05} U \, \Delta T \, dx \qquad \text{for unit length}$$

$$= \frac{2U}{m} \left[N\,e^{mx} - M\,e^{-mx} \right]_{0}^{x=0.05}$$

and substitution of numerical values yields

$$Q_b = 0.684 \text{ kW}$$

The heat that would be transferred *to* the water per unit width if the entire fin area was at the base temperature of 80 °C is

$$Q_f = UA\,\Delta T_b$$

$$= 0.192 \times 2 \times 0.05 \times (80 - 20)$$

$$= 1.152 \text{ kW}$$

The fin efficiency η_f is then

$$\eta_f = \frac{Q_b}{Q_f}$$

$$= \frac{0.684}{1.152}$$

i.e.

$$\eta_f = \underline{59.38\,\%}$$

7.3 Multidimensional Steady-State Conduction

The field of multidimensional steady-state heat transfer has over the years become one of the common testing grounds for mathematical solutions and numerical analyses of the Poisson equation (7.4) and the Laplace equation (7.5). In practice, however, heat transfer problems which may be classed as steady-state conduction and require a precise mathematical solution are seldom encountered. More often the problem is of complicated geometry and the wide tolerances on the available values of the thermal properties nullify any advantages of precision. With this in mind the reader is first introduced to some methods of approximate and rapid determination of the temperature distribution and heat flow, and then to computational methods.

7.3.1 Curvilinear squares and conduction analogues

Flux plotting to form curvilinear squares is a free-hand plotting technique which requires some operator skill but leads to rapid approximate solution of a temperature field. It is applicable to two-dimensional steady-state conduction problems, with simple boundaries and no heat generation. Initially, isotherms are drawn over the field, with an arbitrary but equal temperature increment between each isotherm. Heat flow lines are then drawn to cut each isotherm at right angles since, at each point in the field, heat flows in the direction of the greatest thermal gradient. In addition the heat flow lines are drawn in such a way that they form curvilinear squares with the isotherms; that is 'squares' with curved sides and with equal distances between the mid-points of opposite sides. Referring now to Fig. 7.7, which shows a curvilinear

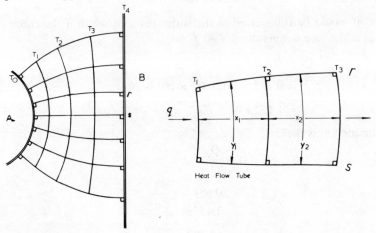

Figure 7.7 Curvilinear squares.

square network plotted for a unit thickness of material, the heat flow q_n along a heat flow tube bounded by heat flow lines r and s is approximately

$$q_n = -ky_1 \frac{(T_2 - T_1)}{x_1} = -ky_2 \frac{(T_3 - T_2)}{x_2}$$

As the squares are curvilinear

$$\frac{x_1}{y_1} = \frac{x_2}{y_2} = 1$$

and for equal temperature increments, ΔT_i

$$T_2 - T_1 = T_3 - T_2 = \Delta T_i$$

Therefore

$$q_n = -k \, \Delta T_i$$

If I is the number of temperature increments between surface A and B (so that $T_B - T_A = I \, \Delta T_i$) and N is the total number of heat flow tubes between A and B, the total heat flow rate Q between A and B is given by

$$Q = -\frac{N}{I} k(T_B - T_A)$$

and for material thickness w

$$Q = -\frac{N}{I} wk(T_B - T_A) \tag{7.16}$$

Example 7.4. A flanged metal duct is surrounded with insulation of conductivity 0.1 W/m K, as in the section shown in Fig. 7.8. The duct and flange

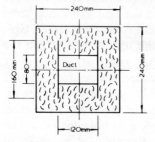

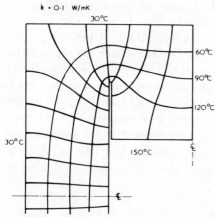

Figure 7.8 Temperature distribution in the insulation around the flanged duct.

are at a temperature of 150 °C and the outer temperature of the insulation is 30 °C. Estimate the approximate heat flow rate per unit length from the duct.

Solution. A free-hand sketch of the curvilinear square network (with 4 temperature increments) for a corner of the insulation is shown in Fig. 7.8. There are $11\frac{1}{2}$ heat flow tubes in the quarter section of insulation shown and the total heat flow per unit length from the duct may be calculated from equation (7.16) as

$$Q = -\left[\frac{11.5}{4} \times 0.1 \times (30 - 150)\right] 4$$

$$= 138 \text{ W/m}$$

This compares favourably with the solution by numerical analysis (Example 7.5) of 149 W/m. With reasonable care the maximum error is unlikely to be greater than one flow tube in the total number of flow tubes plotted which in this case is 1/11.5 or about 9%.

A more precise method of constructing curvilinear squares is available due

127

to the analogy between electrical and thermal flow. The equations due to Laplace (7.5) and Poisson (7.4) may be applied to electrical fields with the temperature term T replaced by the voltage and the heat generation term replaced by an electrical generation term. Electrically conducting paper, typically with a resistance of about 2000 ohms per unit square, is used and the constant voltage (isothermal) boundaries are painted on the paper with silver conducting paint. Equipotential (isothermal) lines are progressively plotted on the paper using a graphite pencil connected to a suitable bridge circuit. Heat flow lines may either be sketched in by hand or preferably by resiting the boundaries and replotting to form a complete orthogonal plot. The arrangement for the flanged duct insulation of Example 7.4 is shown in Fig. 7.9a.

The representation of an infinite field with sources or sinks using this analogy is possible by the following method. Two large circular sheets of conducting paper are placed back to back with a disc of insulating paper between them. Electrical connection is then made between the sheets at the edges either by stapling or covering the edge with silver conducting paint. The shape of the source may then be painted on one side, and connection to the sink at infinity is made at the centre of the other side. The arrangement for constructing the isothermal lines around a Tee-section situated in an infinite field is shown in Fig. 7.9b. Further information on the application of the analogy to open fields is available in Vitovitch and Olsen (1964). An electrical analogy for three-dimensional conduction is possible by using an

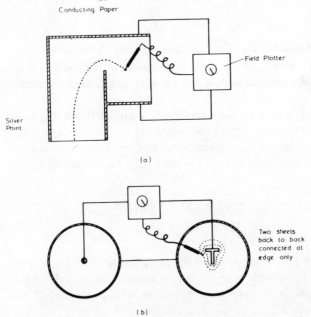

Figure 7.9   Field plotting with conducting paper.

electrolytic tank. A model of the desired shape is positioned in the tank and potentials at required levels are measured using a suitable probe connected to a field plotter, (see Malavard, 1956).

7.3.2 Computational methods

In this sub-section two methods of determining the temperature field are introduced, one involving numerical analysis and the other mathematical computation. Both of these areas are specialized and require an understanding of the particular mathematical concepts involved. In addition the quantity and complexity of the computation necessitates the use of a digital computer for the solution of all but the simpler problems. Here we shall merely consider one or two of these simpler problems and indicate where information on the treatment of more complex conduction problems and computer-aided analysis is available.

Numerical analysis

Consider a homogeneous conducting material of unit thickness with no internal heat source. A simple approach to numerical analysis involves thermally approximating the material to a mesh of conducting rods, as shown in Fig. 7.10, with each rod replacing a width of material equal to the mesh

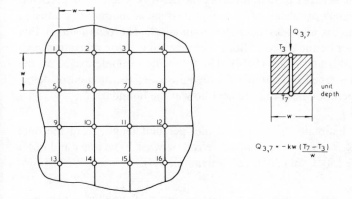

Figure 7.10   Conducting rod mesh.

spacing w. The sum of the heat flows along each rod to an intersection in the mesh is zero under equilibrium conditions and in the case of point 7 for example:

$$Q_{3,7} + Q_{8,7} + Q_{11,7} + Q_{6,7} = 0$$

$$-kw\frac{(T_7 - T_3)}{w} - kw\frac{(T_7 - T_8)}{w} - kw\frac{(T_7 - T_{11})}{w} - kw\frac{(T_7 - T_6)}{w} = 0$$

That is
$$T_3 + T_8 + T_{11} + T_6 - 4T_7 = 0$$

An equation of this type may be obtained for each point in the mesh and the determination of the temperature distribution reduces to the solution of a large number of linear simultaneous equations. These equations may be solved using matrix methods and digital computation. Information on the application of finite difference techniques to heat flow is available in the text by Dusinberre (1961) and suitable Fortran IV computer programs are included in Adams and Rogers (1973) and Welty (1974).

Numerical analysis by hand is possible, and is in fact very suitable for the approximate determination of temperature fields in many practical situations. Furthermore, the ready availability of extremely versatile pocket calculators has considerably reduced the calculation tediousness. The analysis involves initial estimation of mesh point temperatures followed by calculation of the residual R at each point. The residual at point 7 for example is given by

$$T_3 + T_8 + T_{11} + T_6 - 4T_7 = R \tag{7.17}$$

The residual errors at each point are then successively reduced until an acceptable accuracy is achieved and this process is termed relaxation. Careful estimation of the initial temperatures and moderate over-relaxation of points with large residuals can considerably reduce the number of iterations required. Greater accuracy may be obtained by reducing the mesh spacing, over the complete network or in that part of the network where more precision is desired. Practical problems often involve irregular boundaries causing unequal arm lengths at nodes near the boundaries. The treatment of this problem and other boundary conditions is covered in texts on relaxation such as Southwell (1940) and Allen (1954). The following example, based on the insulated duct described in Example 7.4, demonstrates the method of determining the temperature field and the heat flow at the boundaries.

Example 7.5. Estimate the heat flow rate per unit length from the duct illustrated in Fig. 7.8 using the conducting rod network. The inner and outer temperatures are 150 °C and 30 °C (as in Example 7.4).

Solution. A suitable network for a quarter-section of the duct is shown in Fig. 7.11. The initially allotted temperature (estimated by inspection) is included above each point and the corrected temperature after analysis is recorded below. The analysis involves successive calculation and relaxation of the residuals at each point as in Table 7.1. As shown earlier the residual temperature unbalance at a point is the summation of the temperatures of the four surrounding positions less four times the temperature of the point. In this case Table 7.1 shows that point 4 has the largest initial residual of 30 °C and this was reduced to -10 °C by adding 10 °C to point 4:

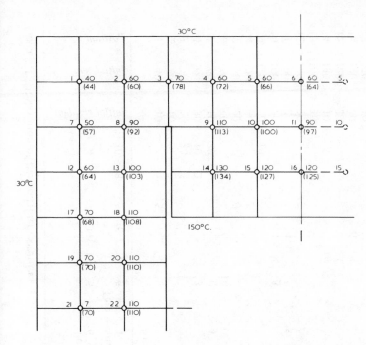

Figure 7.11 Mesh representing the insulation around the flanged duct.

$$T_{side} + T_5 + T_9 + T_3 - 4T_4 = R_4$$

$$30 + 60 + 110 + 70 - 4(70) = -10$$

The residual was over-relaxed to -10 (rather than say $+2$ by adding $7\,°C$ to point 4) because in the initial stages experience shows, that for the overall network, the residuals are reduced in fewer stages by over-relaxation. The effect of the change at point 4 on the surrounding points (3, 5 and 9) was recorded in the table and the new highest residual, point 3, was then relaxed. This process was continued until all the residuals were reduced to $4\,°C$ or less resulting in node temperatures accurate to $1\,°C$. A minor complication occurred at the residual change to point 15 where the effect of adding $6\,°C$ resulted in an alteration of $12\,°C$ to point 16 owing to symmetry; that is at 16,

$$T_{11} + 2T_{15} + T_{duct} - 4T_{16} = R$$

A similar effect occurred at the residual change to point 5 and the second change to point 15.

The heat flow through the insulation may be found by determining the total heat flow from the duct to the nodes or from the nodes to the outer edge or preferably by taking an average of both. The heat flow through the insulation from the duct per unit length (for all four corners) is

TABLE 7.1 Relaxation of the nodal temperatures

Position	Initial temp. °C	Initial residual	Temperature change at position / Residual changes																						Final residual	Final temp. °C
Position relaxed →			4	3	7	11	15	5	16	6	14	1	12	9	13	7	8	18	17	4	5	11	15			
Temperature change at position →			+10	+8	+5	+6	+6	+5	+5	+4	+4	+4	+4	+3	+3	+2	+2	−2	−2	−2	+2	+1	+1			
1	40	10					15					−1	2												1	44
2	60	−10		−2									4												4	60
3	70	20	30	−2						6				6					−2	0			2		0	78
4	60	30	−10	−2							16	−4		8			8			−1	4	1	2		−1	72
5	60	10	20				15		16		−4		0		2	14		2			6	2	1		2	66
6	60	0						0			−14	−8	−3	1	0	4	8		10	2	1	2			2	64
7	50	0				6		5	3			4				2	0		0		2		1		2	57
8	90	20			10					16	−8	1	14		5	0			2	4	4				2	92
9	110	0												14							2	4			4	100
10	100	−20								−14	−4		0		5				−8	−1	0		3		4	113
11	90	20													1						3	1	2		3	97
12	60	10											16			14									−1	64
13	100	10														3			2		4				3	103
14	130	10											−4								4	4	1		−1	127
15	120	20											18						−1		2	1	1		2	134
16	120	20														−2			1		2	1			2	125
17	70	−10																			0	0			0	68
18	70	−10																			1	4	−1		1	108
19	110	−10														−6	−7		−2	−1	−1	0	1		−1	70
20	110	0																			0	−2			−2	110
21	70	0																			−2	−1			0	70
22	110	0																			1	0			0	110

$$Q_{in} = -2k(125-150) - 4k[(127-150) + (134-150) + (134-150)$$
$$+ (113-150) + (78-150) + (92-150) + (103-150) + (108-150)$$
$$+ (110-150)] - 2k(110-150)$$
$$= 153.4 \text{ W/m} \quad \text{(with } k = 0.1 \text{ W/m K)}$$

The heat flow to the surroundings per unit length is

$$Q_{out} = -2k(30-64) - 4k[(30-66) + (30-72) + (30-78) + (30-60)$$
$$+ (30-44) + (30-44) + (30-57) + (30-64) + (30-68) + (30-70)]$$
$$- 2k(30-70)$$
$$= 144 \text{ W/m}$$

The mean heat flow rate is therefore 149 W/m which may be compared with the estimate of 138 W/m by curvilinear squares, (Example 7.4).

Classical mathematical analysis

It was mentioned earlier that some problems involving simple geometries can be solved by classical mathematical methods and solutions are available in Carslaw and Jaeger (1959) and Jakob (1949, 1957). In order to illustrate the approach, the following classical case of a rectangular plate or section with three sides at a constant temperature and the other side at a different temperature is analysed.

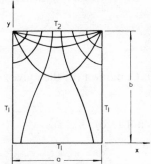

Figure 7.12 The rectangular plate.

The rectangular plate is shown in Fig. 7.12 and the general equation for the temperature distribution in two-dimensions, from Section 7.1, may be given as

$$\frac{\partial^2 \Delta T}{\partial x^2} + \frac{\partial^2 \Delta T}{\partial y^2} = 0 \tag{7.18}$$

where $\Delta T = T - T_1$. This equation may be solved by a method involving separation of the variables. If $X(x)$ is a function of x only and $Y(y)$ is a function of y only, the temperature difference can be expressed as

$$\Delta T = X(x)Y(y)$$

133

Substitution of this expression into equation (7.18) yields on rearrangement

$$-\frac{1}{X}\frac{d^2 X}{dx^2} = \frac{1}{Y}\frac{d^2 Y}{dy^2} \tag{7.19}$$

and since each side is independent as x and y are independent, it follows that each side may be equated to a constant. This separation constant is usually given in the form λ^2, and the following two ordinary differential equations result:

$$\frac{d^2 X}{dx^2} + \lambda^2 X = 0$$

$$\frac{d^2 Y}{dy^2} - \lambda^2 Y = 0$$

The appropriate solutions to these equations are

$$X = A \sin \lambda x + B \cos \lambda x$$

$$Y = C \sinh \lambda y + D \cosh \lambda y$$

and the general solution becomes

$$\Delta T = (A \sin \lambda x + B \cos \lambda x)(C \sinh \lambda y + D \cosh \lambda y) \tag{7.20}$$

The boundary conditions are:

(i) at $x = 0$ $\Delta T = 0$

(ii) at $x = a$ $\Delta T = 0$

(iii) at $y = 0$ $\Delta T = 0$

(iv) at $y = b$ $\Delta T = (T_2 - T_1)$

Application of (i) and (iii) yields $D = 0$ and $B = 0$. Condition (ii) then gives

$$0 = AC \sin \lambda a \sinh \lambda y$$

i.e. $\qquad 0 = \sin \lambda a$ (assuming $\sinh \lambda y$ is not zero)

and $\qquad \lambda = \dfrac{n\pi}{a}$

where $n = 0, 1, 2, 3, \ldots \infty$.

Substitution into equation (7.20) then gives (omitting $n = 0$ which makes no contribution):

$$\Delta T = \sum_{n=1}^{\infty} (AC)_n \sin\left(\frac{n\pi x}{a}\right) \sinh\left(\frac{n\pi y}{a}\right) \tag{7.21}$$

Application of the final boundary condition gives

$$(T_2 - T_1) = \sum_{n=1}^{\infty} (AC)_n \sin\left(\frac{n\pi x}{a}\right) \sinh\left(\frac{n\pi b}{a}\right) \tag{7.22}$$

A solution for the values of $(AC)_n$ may be obtained by comparing this equation with the Fourier series expansion of the constant temperature difference $T_2 - T_1$ between $x = 0$ and $x = a$:

$$(T_2 - T_1) = f(x) = C_1' \sin \lambda_1 x + C_2' \sin \lambda_2 x + C_3' \sin \lambda_3 x \ldots$$

i.e.
$$(T_2 - T_1) = \sum_{n=1}^{\infty} C_n' \sin\left(\frac{n\pi x}{a}\right) \qquad (7.23)$$

where C_n' terms in the Fourier series of period a are given by

$$C_n' = \frac{2}{a} \int_0^a f(x)\sin(\lambda_a x) \, dx$$

$$= \frac{2}{a} \int_0^a (T_2 - T_1)\sin\left(\frac{n\pi x}{a}\right) dx$$

$$= -\frac{2}{a}\left[\frac{(T_2 - T_1)a}{n\pi} \cos\left(\frac{n\pi x}{a}\right)\right]_0^a$$

$$= -2\frac{(T_2 - T_1)}{n\pi}[(-1)^n - 1]$$

Comparison of equations (7.22) and (7.23) shows that

$$(AC)_n \sinh\left(\frac{n\pi b}{a}\right) = C_n' = \frac{2(T_2 - T_1)}{n\pi}[1 - (-1)^n] \qquad (7.24)$$

Substitution of $(AC)_n$ from this equation into equation (7.21) yields the final solution

$$\Delta T = 2(T_2 - T_1) \sum_{n=1}^{\infty} \frac{1 - (-1)^n}{n\pi} \sin\left(\frac{n\pi x}{a}\right) \frac{\sinh(n\pi y/a)}{\sinh(n\pi b/a)} \qquad (7.25)$$

This distribution yields a pattern of isotherms and heat flow lines typically as shown in Fig. 7.12.

7.4 Heat Flow to a Body with Uniform Internal Temperature

Many of the most practical heat conduction problems do not involve the steady heat flow between two thermal reservoirs but rather the transient response to temperature changes. The heating of buildings, the cooking process, foundry casting and the heat treatment of metals by quenching are all cases where the rate of temperature change is important. In this section we shall restrict our study to situations of the type encountered in the final case, that of metal heat treatment, as an example of unsteady-state heat flow to a body of uniform but not constant internal temperature. In particular we shall look at the three systems shown in Fig. 7.13 and consider the thermal change

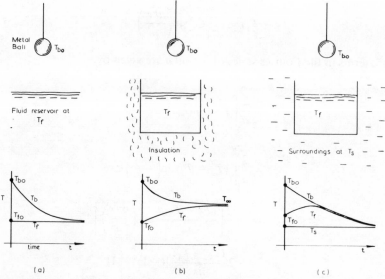

Figure 7.13   Cooling of a metal ball in a fluid.

resulting from the lowering of a metal sphere into a fluid assuming that the temperature within the sphere is uniform. The next section (7.5) is devoted to transient conduction heat transfer within bodies.

Case (a). Situations of the type shown in Fig. 7.13 may be analysed by firstly considering the total heat flow rate from the ball to the fluid and secondly the heat transfer at the surface. In this case a ball is lowered into a large reservoir of fluid at a constant temperature T_f and the heat flow from the ball of mass m_b per unit time t is

$$Q_b = -m_b c_b \frac{dT_b}{dt}$$

and the heat transfer equation is

$$Q_b = -h_b A_b (T_f - T_b)$$

where suffixes b and f refer to the ball or body and the fluid. It therefore follows that

$$\frac{dT_b}{(T_f - T_b)} = \left(\frac{hA}{mc}\right)_b dt$$

and integration with $T_b = T_{bo}$ at $t = 0$ yields

$$\frac{T_b - T_f}{T_{bo} - T_f} = \exp[-(hA/mc)_b t] \qquad (7.26)$$

136

Case (*b*). The ball is lowered into a small quantity m_f of fluid initially at a temperature T_{f0}, which is contained in a well insulated vessel so that all the heat flow from the ball is employed in raising the temperature of the fluid (assuming the heat capacity of the vessel is negligible). The heat flow from the ball is therefore

$$Q_b = -m_b c_b \frac{dT_b}{dt} = m_f c_f \frac{dT_f}{dt} \tag{7.27}$$

and

$$Q_b = -h_b A_b (T_f - T_b) \tag{7.28}$$

A particularly simple solution to these equations is obtained by noting that

$$d(T_f - T_b) = dT_f - dT_b$$

$$= Q_b \left[\frac{1}{m_f c_f} + \frac{1}{m_b c_b} \right] dt$$

Substitution for Q_b from equation (7.28) then gives

$$\frac{d(T_f - T_b)}{(T_f - T_b)} = -h_b A_b \left[\frac{1}{m_f c_f} + \frac{1}{m_b c_b} \right] dt$$

and integration with the boundary conditions $T_f = T_{f0}$ and $T_b = T_{b0}$ at $t = 0$ yields

$$\frac{T_f - T_b}{T_{f0} - T_{b0}} = \exp\left[-h_b A_b \left(\frac{1}{m_f c_f} + \frac{1}{m_b c_b} \right) t \right]$$

Alternatively this may be written as

$$\frac{T_f - T_b}{T_{f0} - T_{b0}} = \exp\left[-H_b \left(\frac{1}{M_f} + \frac{1}{M_b} \right) t \right] \tag{7.29}$$

where the heat capacities $m_f c_f$ and $m_b c_b$ are represented by M_f and M_b and $h_b A_b$ by H_b.

This expression gives the temperature difference between the ball and the fluid at any time t but not the values of T_f and T_b. If these are required it is necessary to solve the simultaneous differential equations (7.27) and (7.28) as follows. Rearrangement of these equations yields:

$$\frac{M_b}{H_b} \frac{dT_b}{dt} = T_f - T_b \tag{7.30}$$

$$-\frac{M_f}{H_b} \frac{dT_f}{dt} = T_f - T_b \tag{7.31}$$

Substitution of T_b from (7.31) into (7.30) gives

$$\frac{M_b}{H_b} \left[\frac{dT_f}{dt} + \frac{M_f}{H_b} \frac{d^2 T_f}{dt^2} \right] = T_f - \left[T_f + \frac{M_f}{H_b} \frac{dT_f}{dt} \right]$$

$$\frac{d^2 T_f}{dt^2} = -H_b \left[\frac{M_b + M_f}{M_b M_f} \right] \frac{dT_f}{dt}$$

$$= -H_b \left[\frac{1}{M_f} + \frac{1}{M_b} \right] \frac{dT_f}{dt}$$

which has a general solution of the form

$$T_f = A \exp \left[-H_b \left(\frac{1}{M_f} + \frac{1}{M_b} \right) t \right] + B$$

With the boundary conditions $T_f = T_{f0}$ at $t = 0$ and $T_f = T_\infty$ at $t = \infty$ (see Fig. 7.13) this becomes

$$\frac{T_f - T_\infty}{T_{f0} - T_\infty} = \exp \left[-H_b \left(\frac{1}{M_f} + \frac{1}{M_b} \right) t \right] \tag{7.32}$$

Similarly it can be shown that

$$\frac{T_b - T_\infty}{T_{b0} - T_\infty} = \exp \left[-H_b \left(\frac{1}{M_f} + \frac{1}{M_b} \right) t \right] \tag{7.33}$$

The value of T_∞ may be found from the overall energy conservation:

$$M_b T_{b0} + M_f T_{f0} = T_\infty (M_b + M_f)$$

from which

$$T_\infty = \frac{M_b T_{b0} + M_f T_{f0}}{M_b + M_f} \tag{7.34}$$

Finally equations (7.32) and (7.33) may be written as

$$\frac{T_f - T_\infty}{T_{f0} - T_\infty} = \frac{T_b - T_\infty}{T_{b0} - T_\infty} = K$$

where K represents the exponential term. Expression of T_f and T_b in terms of K and subtraction yields

$$T_f - T_b = K(T_{f0} - T_{b0})$$

which is equivalent to equation (7.29).

Case (c). The ball is lowered into a small quantity of fluid, initially at a temperature T_{f0}, contained in a vessel which is open to surroundings at a constant temperature T_s. As in the previous cases the heat flow Q_b from the ball is given by

$$Q_b = -m_b c_b \frac{dT_b}{dt}$$

$$Q_b = -h_b A_b (T_f - T_b)$$

Heat flow to the surroundings is given by

$$Q_s = -h_s A_s (T_s - T_f)$$

where h_s and A_s refer to the mean heat transfer coefficient between the vessel and the surrounding air and the effective vessel area. Assuming the heat capacity of the vessel is small the net heat flow to the fluid is

$$Q_b - Q_s = m_f c_f \frac{dT_f}{dt}$$

Rearrangement of these equations using $M = mc$ and $H = hA$ as previously yields

$$M_b \frac{dT_b}{dt} = H_b (T_f - T_b)$$

$$M_f \frac{dT_f}{dt} = -H_b (T_f - T_b) + H_s (T_s - T_f)$$

The analysis of these non-homogeneous simultaneous differential equations is more easily performed for a particular case than for the general case but the method is as follows. In operator form the equations become

$$M_b D(T_b) = H_b (T_f - T_b) \tag{7.35}$$

$$M_f D(T_f) = -H_b (T_f - T_b) + H_f (T_s - T_f) \tag{7.36}$$

From equation (7.36)

$$(M_f D + H_b + H_f) T_f = H_b T_b + H_f T_s$$

Substitution for T_f in equation (7.35) and rearrangement yields

$$[M_b M_f D^2 + (M_b H_b + M_b H_f + M_f H_b) D + H_f H_b] T_b = H_f H_b T_s$$

which has a solution of the form

$$T_b = A\, e^{mt} + B\, e^{-mt} + T_s \tag{7.37}$$

where $m = \left[\dfrac{(M_b H_b + M_b H_f + M_f H_b)^2 - 4 M_b M_f H_f H_b}{2 M_b M_f} \right]^{\frac{1}{2}}$

and A and B are constants determined by the boundary conditions. The equation for T_f is found in a similar way by substituting the expression for T_b from (7.35) into (7.36).

Example 7.6. In a coating process toy cars constructed of steel sheet are immersed in a bath of molten plastic at a temperature of 200 °C and then allowed to cool in ambient air at a temperature of 20 °C. The cars have a mass of 0.1 kg and a total surface area of 0.06 m^2 and the specific heat of the steel is 0.46 kJ/kg K. The plastic layer on the cars has a mean thickness of 0.3 mm, a thermal conductivity of 0.2 W/m K and sets at 130 °C. The natural convection heat transfer coefficient between the car and the surrounding air is given approximately by

$$h = 0.0013\left(\frac{\Delta T}{0.1}\right)^{0.25} \text{ kW/m}^2 \text{ K}$$

(from Table 3.3 with $d = 0.1$ m). Estimate the time that should be allowed for the plastic coat to set after withdrawal of the cars from the bath (neglecting radiation and the mass of the plastic on the car).

Solution. The equations for heat flow from the car to the surroundings are

$$Q = -mc\frac{dT_c}{dt} \tag{7.38}$$

$$Q = -UA(T_a - T_c) \tag{7.39}$$

where m, c and A refer to the mass, specific heat and surface area of the car, T_a and T_c are the air and car temperatures and U is the overall heat transfer coefficient from the car, given by

$$\frac{1}{U} = \frac{x}{k} + \frac{1}{h}$$

$$= \frac{0.3 \times 10^{-3}}{0.2 \times 10^{-3}} + \frac{1}{0.0013\left(\dfrac{\Delta T}{0.1}\right)^{0.25}} \frac{\text{m}^2 \text{ K}}{\text{kW}}$$

In this case $\Delta T = T_c - T_a$ and varies from an initial 180 °C to 110 °C. Numerical substitution shows that the x/k term is negligible compared to the $1/h$ term and equation (7.39) becomes

$$Q = +\frac{0.0013A}{(0.1)^{0.25}}(T_c - T_a)^{1.25}$$

$$= 0.00231 \, A(T_c - T_a)^{1.25}$$

Elimination of Q using equation (7.38) then yields

$$\frac{dT_c}{(T_c - T_a)^{1.25}} = \frac{-0.00231A}{mc} dt$$

and

$$-4(T_c - T_a)^{-0.25} = -\frac{0.00231A}{mc}t + C$$

Substitution of $(T_c - T_a) = 180$ °C at $t = 0$ gives $C = -1.092$. It is required to find t when $(T_c - T_a) = 110$ °C:

$$t = [-1.092 + 4(110)^{-0.25}] \frac{0.1 \times 0.46}{0.00231 \times 0.06}$$

$$t = 47.5 \text{ s}$$

A setting time of 1 minute is therefore ample.

Finally it is once again emphasized that this approach to unsteady-state situations (sometimes called the lumped-heat-capacity approach) is only applicable where the temperature variation within the body is small and the main thermal resistance is at the surface. The case of metal objects cooling in a surrounding fluid is one of the more common applications but it is equally applicable to non-metallic structures with a large surface per unit volume.

7.5 Unsteady-State Conduction

Mathematical analysis

The unsteady-state heat conduction equation as derived in Section 7.1 may be written in the following form for uniform conductivity and no internal heat generation

$$\frac{\partial^2 T}{dx^2} + \frac{\partial^2 T}{\partial y^2} + \frac{\partial^2 T}{dz^2} = \frac{1}{\alpha} \frac{\partial T}{\partial t} \tag{7.40}$$

Mathematical solutions of this equation and its equivalent in cylindrical and spherical coordinates are complex even for conduction in one dimension. The solution for the case of the temperature distribution in a semi-infinite plate which has undergone a step temperature change at the surface is given by

$$\frac{T - T_s}{T_i - T_s} = \text{erf}\left(\frac{x}{2\sqrt{\alpha t}}\right) \tag{7.41}$$

where the error function is

$$\text{erf}\left(\frac{x}{2\sqrt{\alpha t}}\right) = \frac{2}{\sqrt{\pi}} \int_0^{x/(2\sqrt{\alpha t})} e^{-\eta^2} \, d\eta$$

In this equation T is the temperature at time t and distance x from the surface, T_s is the constant temperature at the surface after the change, T_i is the initial uniform temperature of the plate and η is a dummy variable.

For simple geometrical arrangements involving one-dimensional unsteady conduction, charts giving temperature variation as a function of time are available. The parameters are usually represented in dimensionless form as follows, where d is a physical dimension and k is the conductivity of the solid:

Biot Number $\text{Bi} = \dfrac{hd}{k}$

Fourier Number $\text{Fo} = \dfrac{\alpha t}{d^2}$

Dimensionless temperature difference $\theta = \dfrac{T - T_f}{T_i - T_f}$ (in fluid surroundings)

The general relationship for a body initially at T_i immersed in an infinite reservoir of fluid at T_f is

$$\theta = f(\text{Bi, Fo})$$

Equation (7.41) is the particular form of this relationship for a semi-infinite plate when h and Bi tend to infinity so that T_s tends to T_f. It may be rewritten as

$$\theta = \text{erf}\left(\frac{1}{2\sqrt{\text{Fo}_x}}\right)$$

A chart giving the centre temperature of a sphere in a fluid of known temperature is shown in Fig. 7.14 and the following example illustrates its use.

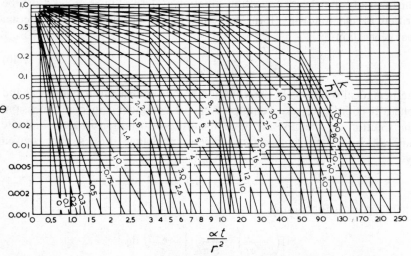

Figure 7.14   Centre temperature of a sphere. (Adapted from Heisler, 1947).

Example 7.7. An egg with a mean diameter of 40 mm and initially at a temperature of 20 °C is placed in a saucepan of boiling water for 4 minutes and found to be boiled to the consumer's taste. For how long should a similar egg for the same consumer be boiled when taken from a refrigerator at a temperature of 5 °C? The egg properties are $k = 2$ W/m K, $\rho = 1200$ kg/m³, $c = 2$ kJ/kg K and the heat transfer coefficient for the shell and shell-water interface may be taken as 0.2 kW/m² K. Compare the centre temperature attained with that computed by treating the egg as a lumped-heat-capacity system.

142

Solution. We shall base our solution on the assumption that the attainment of a certain centre temperature is sufficient to define the condition of the boiled egg. The centre temperature after 4 minutes is found as follows (where the physical dimension is taken as the radius in this case to enable direct use of the chart).

$$\text{Bi}_r = \frac{hr}{k} = \frac{0.2 \times 0.02}{2 \times 10^{-3}} = 2$$

$$\text{Fo}_r = \frac{kt}{\rho c r^2} = \frac{2 \times 10^{-3} \times 4 \times 60}{1200 \times 2 \times 4 \times 10^{-4}} = 0.5$$

From the chart, with $k/hr = 1/\text{Bi}_r = 0.5$,

$$\theta = 0.20 = \frac{T - 100}{20 - 100}$$

$$T = 84\,°\text{C}$$

When the egg is taken from the refrigerator the new value of θ is

$$\theta = \frac{84 - 100}{5 - 100} = 0.168$$

From the chart (with the same value of k/hr) $\text{Fo}_r \approx 0.55$ and the new value of t is given by

$$\frac{t}{4 \times 60} = \frac{0.55}{0.5}$$

$$t = 264 \text{ seconds}$$

and the egg should therefore be boiled 24 seconds longer.

Analysing the heat flow to the egg, initially at $20\,°\text{C}$, as a lumped-heat-capacity system, equation (7.26) yields

$$\frac{T - 100}{20 - 100} = \exp\left[\frac{-0.2\pi(0.04)^2 \times 4 \times 60}{2 \times 1200\pi(0.04)^3/6}\right]$$

$$T = 96\,°\text{C}$$

The discrepancy is mainly due to the assumption of uniform temperature throughout the egg, an assumption which is necessary for application of the lumped-heat-capacity method.

Graphical analysis

A convenient method of determining the approximate temperature distribution in the more complicated geometrical arrangements generally encountered in engineering practice is provided by the Schmidt graphical technique (after Schmidt, 1936). This technique is suitable for applications involving various surface conditions and is capable of accommodating changes of conductivity and surface heat transfer coefficient with temperature. The brief description included here is sufficient for the analysis of many common problems and for a more rigorous treatment the reader is referred to Jakob (1949, 1957). The general field of numerical analysis as applied to unsteady conduction is covered in the texts of Dusinberre (1961) and, in conjunction with Fortran programming, by Adams and Rogers (1973).

From equation (7.40) the heat conduction equation in one dimension is

$$\frac{\partial^2 T}{\partial x^2} = \frac{1}{\alpha}\frac{\partial T}{\partial t} \tag{7.42}$$

and this may be written in finite difference form as

$$\frac{\Delta_x(\Delta_x T/\Delta x)}{\Delta x} = \frac{\Delta_t T}{\propto \Delta t} \tag{7.43}$$

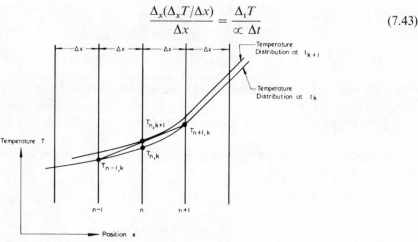

Figure 7.15 The graphical method.

Reference to Fig. 7.15 giving the notation associated with the temperature distribution in a material at a time t_k and one increment later at a time t_{k+1} indicates that, at position n,

$$\Delta_t T = T_{n,\,k+1} - T_{n,\,k}$$

and at time t_k

$$\Delta_x T = T_{n+1,\,k} - T_{n,\,k}$$

(Note that suffix n refers to position and suffix k to time.) The numerator of the left-hand side of equation (7.43) represents the change of slope of $\Delta_x T/\Delta x$ with respect to Δx and is therefore dependent on the difference between two successive values of $\Delta_x T$:

$$\Delta_x\left(\frac{\Delta_x T}{\Delta x}\right) = \left(\frac{T_{n+1,\,k} - T_{n,\,k}}{\Delta x}\right) - \left(\frac{T_{n,\,k} - T_{n-1,\,k}}{\Delta x}\right)$$

$$= \frac{T_{n+1,\,k} - 2T_{n,\,k} - T_{n-1,\,k}}{\Delta x}$$

144

Substitution in equation (7.43) and rearrangement then yields

$$T_{n,k+1} - T_{n,k} = \frac{\alpha \, \Delta t}{(\Delta x)^2}(T_{n+1,k} - 2T_{n,k} - T_{n-1,k}) \tag{7.44}$$

If the time increment is selected such that a straight line joining points $T_{n-1,k}$ and $T_{n+1,k}$ passes through the point $T_{n,k+1}$ (as in Fig. 7.15) the geometrical arrangement yields

$$T_{n,k+1} = \tfrac{1}{2}(T_{n+1,k} - T_{n-1,k})$$

and

$$T_{n,k+1} - T_{n,k} = \tfrac{1}{2}(T_{n+1,k} - 2T_{n,k} - T_{n-1,k}) \tag{7.45}$$

Comparison of equations (7.44) and (7.45) shows that under conditions when

$$\Delta t = \frac{(\Delta x)^2}{2\alpha} \tag{7.46}$$

the temperature change at point $T_{n,k}$ after a time increment Δt is indicated by the intercept obtained on constructing a straight line between points spaced Δx each side on the temperature distribution plot. The following example may serve to clarify the procedure.

Example 7.8. A wall of insulation material with a thermal diffusivity of 0.333×10^{-6} m^2/s has a thickness of 50 mm and is initially at a uniform temperature of 20 °C. The right-hand side of the wall is suddenly increased to a constant temperature of 28 °C and at the same time the temperature of the left-hand side is gradually increased at the rate of 4 °C per minute. Determine the time elapse to the point when all the wall is at a temperature of more than 28 °C.

Solution. A graphical solution is shown in Fig. 7.16. Either Δt or Δx is arbitrarily selected, (as they are not independent owing to the required interrelationship of equation (7.46)). In this case a Δx of 1 cm was selected as it was a convenient subdivision of the wall thickness and because 1 cm square graph paper was to hand. The initial temperature distribution was uniform at 20 °C and after time Δt, calculated as $2\frac{1}{2}$ minutes, the right-hand side of the distribution was at 28 °C as specified, and the left-hand side had increased to 30 °C this being an increase of 4 °C per minute. Temperature distributions at increments of $2\frac{1}{2}$ minutes were progressively constructed by connecting points at $2(\Delta x)$ spacing on the previous curve as shown in Fig. 7.15. It was found that the lowest temperature exceeded 28 °C after 15 minutes. A more precise distribution at any time could be obtained by drawing a continuous curve tangential to the centre of each linear section.

Surface boundary conditions involving a fluid and a known heat transfer coefficient may be accommodated in the graphical technique by determining

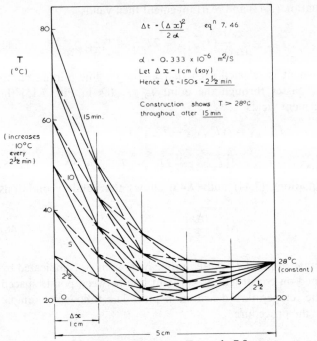

Figure 7.16 Graphical solution to Example 7.8.

the slope of the temperature distribution on the fluid side of the surface. At the surface the conduction and convection heat fluxes are equal and

$$q = -k\frac{\mathrm{d}T}{\mathrm{d}x} = -h(T_f - T_b)$$

where suffixes f and b denote fluid and boundary. The slope of the temperature distribution is given by

$$\text{slope} = \frac{\mathrm{d}T}{\mathrm{d}x} = \frac{(T_f - T_b)}{k/h}$$

and in the construction this means that the tangent of the temperature profile at the surface passes through the fluid temperature at a distance k/h from the surface. Since the centre of each increment Δx of the graphically obtained temperature distribution has the most precise slope, it is advisable to arrange the increments so that the surface falls in the centre. The following example illustrates the application of the fluid boundary technique together with the treatment of constant heat flux boundaries. Many surface conditions involving non-linear temperature changes and variation of h with temperature may be solved by using ingenuity in the graphical construction.

Example 7.9. The side of a large industrial 'off-peak' storage heater consists essentially of a wall of insulation brick 120 mm thick. The inner side receives

146

a steady flow of heat at the rate of 0.7 kW/m^2 and the outer side is in a steady air temperature of $20\,°\text{C}$. The heat transfer coefficients for the inner and outer sides are 0.012 and $0.010 \text{ kW/m}^2 \text{ K}$ and the thermal diffusivity of the insulation is $0.333 \times 10^{-6} \text{ m}^2/\text{s}$. The temperature of the air on the inner side when the steady-state condition is reached is $400\,°\text{C}$.

Estimate the surface temperatures of the insulation at times of 20 minutes and 4 hours after the wall first starts heating up from a temperature of $20\,°\text{C}$. What is the heat storage rate after 4 hours?

Solution. The thermal conductivity of the wall is not given and must be determined from information on the steady-state condition. From:

and
$$q = -U(T_{ao} - T_{ai})$$

$$\frac{1}{U} = \frac{1}{h_i} + \frac{x}{k} + \frac{1}{h_o}$$

(where suffixes i and o denote inner and outer sides and T_{ai} and T_{ao} indicate the temperature of the air on inner and outer sides) it follows that

$$\frac{x}{k} = -\frac{(T_{ao} - T_{ai})}{q} - \frac{1}{h_i} - \frac{1}{h_o}$$

$$= \frac{380}{0.7} - \frac{1}{0.012} - \frac{1}{0.010}$$

$$= 360$$

and

$$k = \frac{0.12}{360} = 0.333 \times 10^{-3} \text{ kW/m K}$$

The inner and outer k/h distances are therefore

$$\frac{k}{h_i} = \frac{0.333 \times 10^{-3}}{0.012} = 0.028 \text{ m} = 2.8 \text{ cm}$$

$$\frac{k}{h_o} = \frac{0.333 \times 10^{-3}}{0.010} = 0.033 \text{ m} = 3.3 \text{ cm}$$

The construction is shown in Fig. 7.17 and the initial temperature distribution is taken at the point where the wall is about to be heated but the air on the inner side is already warm. On the inner side the heat flux is constant at 0.7 kW/m^2 and as $q \propto \mathrm{d}T/\mathrm{d}x$ the slope of the temperature distribution is constant and equal to the slope in the steady-state condition. The construction therefore also yields the value of the air temperature as a function of time. On the outer side the air temperature is constant and all the temperature plots radiate from the same point. Initially an increment of $\Delta x = 2$ cm is selected leading to a time increment Δt of 10 minutes, and after 20 minutes the required surface temperatures are $67\,°\text{C}$ on the inner side and no change on the outer side. After an interval of 40 minutes Δx is increased to 4 cm and Δt

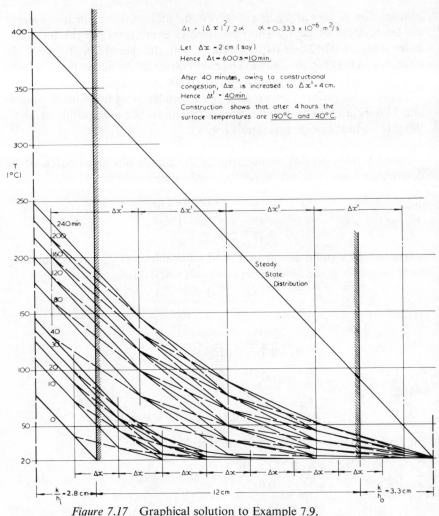

Within the figure:

$\Delta t \cdot (\Delta x)^2 / 2\alpha \qquad \alpha = 0.333 \times 10^{-6} \ m^2/s$

Let $\Delta x = 2$ cm (say)
Hence $\Delta t = 600 s = 10$ min.

After 40 minutes, owing to constructional
congestion, Δx is increased to $\Delta x' = 4$ cm.
Hence $\Delta t' = 40$ min.
Construction shows that after 4 hours the
surface temperatures are $\underline{190°C}$ and $\underline{40°C}$.

Figure 7.17 Graphical solution to Example 7.9.

to 40 minutes and further plotting (commencing with the temperature
distribution at 40 minutes) yields surface temperatures of 190 °C and 40 °C
after 4 hours.

The heat storage rate after 4 hours is given by the difference between the
heat input and output rates at this time. The input rate is given as 0.7 kW/m^2
and the output (on the outer side) is given by $q = -k \ dT/dx$ where the gradient is determined from the construction:

i.e.,

$$q_{store} = q_{in} - q_{out}$$

$$= 0.7 + 0.333 \times 10^{-3} \times \frac{-20 \ (°C)}{3.3 \ (cm) \times 10^{-2}}$$

$$= 0.5 \ kW/m^2 \ .$$

148

Forced Convection Analysis

8.1 Introduction

The heat flow rate between a moving fluid and a wall is controlled by a thin layer of fluid at the surface, in which the flow is restricted. The flow velocity varies from the free-stream value for the bulk of the fluid to zero at the surface and the region in which this occurs is termed the boundary layer. Some qualitative aspects of the boundary layer were discussed in Chapter 3 and in this chapter we shall attempt to demonstrate how a theoretical analysis can lead to relationships which predict the heat flow rate through the boundary layer.

The conventional approach in the examination of fluid motion is to consider a certain well-defined part of the field of flow termed the control volume and to make balances of the flow of mass, momentum and energy through the control volume. This approach leads to the generalized continuity, momentum and energy equations. The equations may be expressed in differential or integral form, in Cartesian, radial or vector coordinates and in one or more dimensions. Furthermore, the continuity and momentum equations may be combined to give the Navier–Stokes equations, which may also exist in a number of forms. Fortunately the situation is considerably simplified when considering the boundary layer owing to the following restrictions and assumptions which may generally be applied.

1. The analysis is restricted to steady flow. This strictly implies laminar flow in the boundary layer, although the analysis is often applied to turbulent boundary layers.

2. The boundary layer is assumed to be thin so that in forced convection the velocity along the layer is much greater than the velocity across the layer. In the nomenclature shown in Fig. 8.1, $u \gg v$.

3. The rate of change of velocity u across the layer is much greater than the rate of change of the velocity along the layer. That is

$$\frac{\partial u}{\partial y} \gg \frac{\partial u}{\partial x}.$$

4. The thermal boundary layer is also thin and similarly to 3,

$$\frac{\partial T}{\partial y} \gg \frac{\partial T}{\partial x}$$

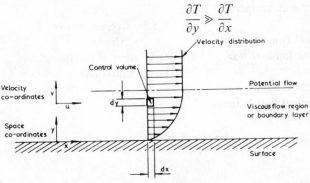

Figure 8.1 Boundary layer coordinates.

5. The fluid property changes with temperature and position are assumed to be sufficiently small (except in special cases) to enable constant values to be used.

6. The variation of fluid pressure in the direction of flow is assumed to be negligible.

7. Potential flow is assumed to exist outside the boundary layer.

8. Body forces acting on the fluid caused by gravity and magnetic fields are assumed to have a negligible effect on the analysis.

9. Terms involving viscous energy dissipation are also assumed to have a negligible effect. Viscous energy, which arises due to friction within the fluid, is dissipated as heat and at normal velocities this internal heat generation is very small. At high velocities this effect must be included.

With these simplifications in mind the differential continuity, momentum and energy equations of the boundary layer together with an integrated form of the energy equation may be expressed in the following form:

Mass continuity

$$\frac{\partial u}{\partial x} + \frac{\partial v}{\partial y} = 0 \qquad (8.1)$$

Momentum

$$u \frac{\partial u}{\partial x} + v \frac{\partial u}{\partial y} = v \frac{\partial^2 u}{\partial y^2} \qquad (8.2)$$

Energy

$$u \frac{\partial T}{\partial x} + v \frac{\partial T}{\partial y} = \alpha \frac{\partial^2 T}{\partial y^2} \qquad (8.3)$$

Heat flow equation (an integrated form of the energy equation)

$$\frac{d}{dx} \left(\int_0^Y (T_\infty - T)u \, dy \right) = \alpha \left(\frac{dT}{dy} \right)_0 \qquad (8.4)$$

Each of these equations is derived in Section 8.4. In the following two sections we shall cover geometrical arrangements involving both internal and external boundary layers and shall make use of these equations together with empirical relationships. We shall also solve some convection heat transfer problems to illustrate the application of the analysis. In general, laminar flow is more suitable for analytical study, although in recent years an increasing number of turbulent flow problems are being provided with theoretical solutions; see for example Spalding and Patankar (1967). (In connection with Section 3.2 it should be noted that, under similar flow conditions, the velocity and temperature distributions indicated by equations (8.2) and (8.3) are similar, and have exactly the same form when the criterion for Reynolds analogy is fulfilled; that is when $v/\alpha = \mathrm{Pr} = 1$.)

8.2 Flow Inside Tubes

8.2.1 The energy equation for laminar flow in tubes

For analysis of axial flow in tubes the energy equation (8.3) may be transformed into cylindrical radial coordinates. When the flow is laminar and fully developed the radial velocity is zero, and with axial symmetry there is no variation of the temperature with radial angle θ. Under these conditions the energy equation is

$$\frac{1}{r}\frac{\partial}{\partial r}\left(r\frac{\partial T}{\partial r}\right) = \frac{u}{\alpha}\frac{\partial T}{\partial x} \qquad (8.5)$$

Alternatively this equation may be simply derived by equating the heat conducted radially and the heat convected axially in an annular control volume within the flow.

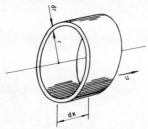

Figure 8.2 Annular control volume.

within the flow. Referring to Fig. 8.2 the radial heat flow q_r into the control volume is given by

$$dq_{r,\,\text{in}} = -k2\pi r\,dx\,\frac{\partial T}{\partial r}$$

and out of the control volume is

$$dq_{r,\,\text{out}} = -k2\pi(r+dr)\,dx\left[\frac{\partial T}{\partial r} + \frac{\partial}{\partial r}\left(\frac{\partial T}{\partial r}\right)dr\right]$$

The net heat flow out of the control volume by axial convection in the x-direction is

$$dq_{\text{conv}} = d\dot{m}c_p\left(\frac{\partial T}{\partial x}\right)dx$$

$$= \rho u 2\pi r\,dr c_p\left(\frac{\partial T}{\partial x}\right)dx$$

A heat balance neglecting small quantities then yields

$$\frac{1}{r}\left(\frac{\partial T}{\partial r} + r\frac{\partial^2 T}{\partial r^2}\right) = \left(\frac{\rho c_p}{k}\right)u\frac{\partial T}{\partial x}$$

which is the same as equation (8.5).

8.2.2 The laminar flow velocity profile

In order to solve equation (8.5) for the temperature distribution and ultimately for the heat flow rate to a fluid in a tube it is necessary to determine the velocity distribution u in terms of r. This may be achieved by considering the fluid dynamics of the fully developed flow. A balance between the pressure

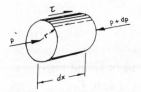

Figure 8.3 Fluid element in a tube.

and shear forces on a concentric cylinder of fluid in a tube as shown in Fig. 8.3 yields

$$\pi r^2(p + dp) - \pi r^2 p = 2\pi r\tau\, dx$$

Substitution of $\tau = \mu(du/dr)$ and rearrangement gives

$$du = \frac{r}{2\mu}\frac{dp}{dx}\, dr$$

and

$$u = \frac{r^2}{4\mu}\frac{dp}{dx} + \text{constant}$$

At the wall of the tube $u = 0$, $r = r_0$ and therefore

$$u = \frac{(r^2 - r_0^2)}{4\mu}\frac{dp}{dx}$$

It is convenient to express u in terms of the bulk or mean velocity U generally associated with fluid flow. (This bulk velocity is the value yielded by using the continuity equation in the form $U = \dot{m}/\rho A_c$ and is similar to the bulk mean temperature discussed in Section 5.2.) It is defined as

$$U = \frac{\displaystyle\int_{A_c} u\, dA_c}{A_c} \tag{8.6}$$

where A_c is the cross-sectional area.

For a circular tube this becomes

$$U = \frac{\displaystyle\int_0^{r_o} u 2\pi r\, dr}{\pi r_0^2}$$

and on substitution for u

$$U = \frac{1}{2\mu}\left(\frac{dp}{dx}\right)\int_0^{r_o} \frac{(r^2 - r_0^2)r}{r_0^2}\, dr$$

$$= -\frac{r_0^2}{8\mu}\left(\frac{dp}{dx}\right)$$

and

$$u = 2U\left(1 - \frac{r^2}{r_0^2}\right) \tag{8.7}$$

Before we substitute this back into the energy equation (8.5) and attempt a solution it is necessary to have a closer look at the temperature T.

It is convenient to express T in terms of the bulk or mean temperature T_m of the flow as defined by equation (5.2).

For the case of a tube this becomes

$$T_m = \frac{1}{\pi r_0^2 U} \int_0^{r_0} uT(2\pi r)\, dr$$

$$= \frac{2}{r_0^2 U} \int_0^{r_0} uTr\, dr \tag{8.8}$$

Furthermore it is also convenient to define a non-dimensional temperature difference θ, which is zero at the wall and maximum at the axis of the tube, in the form

$$\theta = \frac{T_0 - T}{T_0 - T_m} \tag{8.9}$$

Under conditions when the temperature profile is fully developed θ is assumed to be invariant along the tube and the heat transfer coefficient is constant (as discussed in Kays 1966, Chapter 8). The constant temperature profile means that

$$\frac{\partial}{\partial x}\left(\frac{T_0 - T}{T_0 - T_m}\right) = 0 \tag{8.10}$$

For a complete solution to the energy equation the thermal boundary condition at the tube wall must be specified. There are two situations commonly encountered for which simple mathematical solutions are possible. One is for a constant heat flow rate at the wall (as may occur in electrical or radiant heating) and the other is for a constant wall temperature. We shall consider each of these in turn.

8.2.3 Solution of the laminar flow energy equation for constant heat flux at the wall

In tube flow the heat transfer coefficient is defined in terms of the bulk temperature T_m and the wall temperature T_0 by

$$q = -h(T_m - T_0)$$

where q is the heat flux from the wall to the fluid.

Since in this case q and h are constant it follows that

$$(T_m - T_0) = \text{constant}$$

and

$$\frac{\partial T_m}{\partial x} - \frac{\partial T_0}{\partial x} = 0$$

Also from equation (8.10) when $(T_0 - T_m)$ is constant

$$\frac{\partial T_0}{\partial x} - \frac{\partial T}{\partial x} = 0$$

and therefore

$$\frac{\partial T}{\partial x} = \frac{\partial T_m}{\partial x} \tag{8.11}$$

Substitution of the velocity profile, equation (8.7), and also equation (8.11) into the energy equation (8.5) yields:

$$\frac{1}{r}\frac{\partial}{\partial r}\left(r\frac{\partial T}{\partial r}\right) = \frac{2U}{\alpha}\left(1 - \frac{r^2}{r_0^2}\right)\frac{\partial T_m}{\partial x}$$

Integration gives

$$\frac{r\,\partial T}{\partial r} = \frac{2U}{\alpha}\left(\frac{r^2}{2} - \frac{r^4}{4r_0^2}\right)\frac{\partial T_m}{\partial x} + C_1$$

and

$$T = \frac{2U}{\alpha}\left(\frac{r^2}{4} - \frac{r^4}{16r_0^2}\right)\frac{\partial T_m}{\partial x} + C_1 \ln r + C_2$$

Substitution of the boundary conditions

$$T = T_0 \text{ at } r = r_0$$

and

$$\frac{\partial T}{\partial r} = 0 \quad \text{at } r = 0$$

yields:

$$T = T_0 - \frac{2U}{\alpha}\left(\frac{3}{16}r_0^2 + \frac{r^4}{16r_0^2} - \frac{r^2}{4}\right)\frac{\partial T_m}{\partial x} \tag{8.12}$$

Substitution of equations (8.7) and (8.12) for u and T in equation (8.8), and rearrangement gives:

$$T_m = \int_0^{r_0} \frac{4}{r_0^2}\left(1 - \frac{r^2}{r_0^2}\right)\left[T_0 - \frac{2U}{\alpha}\left(\frac{3}{16}r_0^2 + \frac{r^4}{16r_0^2} - \frac{r^2}{4}\right)\frac{\partial T_m}{\partial x}\right]r\,dr$$

$$= T_0 - \frac{11}{48}\frac{Ur_0^2}{\alpha}\frac{\partial T_m}{\partial x}$$

The heat flux is then

$$q = -h(T_m - T_0)$$

$$= h\left(\frac{11}{48} \frac{Ur_0^2}{\alpha} \frac{\partial T_m}{\partial x}\right) \qquad (8.13)$$

The heat flux is also given by the temperature gradient in the fluid at the wall:

$$q = -k\left(\frac{\partial T}{\partial r}\right)_{r=r_0}$$

By differentiating equation (8.12) with respect to r and substituting $r = r_0$

$$q = k \frac{2U}{\alpha} \frac{r_0}{4} \frac{\partial T_m}{\partial x} \qquad (8.14)$$

Comparing equations (8.13) and (8.14),

$$h = \frac{24}{11} \frac{k}{r_0}$$

In terms of the Nusselt number, with $d = 2r_0$,

$$\mathrm{Nu} = \frac{hd}{k} = \frac{24}{11} \times 2$$

$$\mathrm{Nu} = 4.364 \qquad (8.15)$$

and the heat transfer coefficient for this situation is therefore a function of the tube diameter and the fluid conductivity only.

Example 8.1. Hot air flowing through a metal pipe of 20 mm diameter is cooled at a constant rate per unit length of pipe. At a particular section, denoted ⓐ, the air velocity in the centre of the pipe is found to be 2 m/s and the wall temperature, as measured by a thermocouple on the inner surface of the pipe, is 250 °C. At a section, denoted ⓑ, situated 1 m downstream from ⓐ, the wall temperature is found to be 200 °C. Estimate the mean air temperature at section ⓑ.

Solution. An indication of whether the flow is laminar or turbulent is given by the Reynolds number, $\mathrm{Re} = \rho U d / \mu$, where U is the mean or bulk velocity in tube flow. Assuming for the moment that the flow is laminar the bulk velocity may be determined from the measured centre-line velocity by equation (8.7):

$$u = 2U(1 - r^2/r_0^2)$$

and at $r = 0$, $u = 2U$. The bulk velocity is therefore 1 m/s and the Reynolds number may be estimated (substituting property values for air at 500 °K) as

$$\mathrm{Re} = \frac{0.706 \times 1 \times 0.02}{2.67 \times 10^{-5}} = 528$$

thus confirming (as Re is less than Re_{crit} of about 2300) that the flow is laminar. The mean air temperature at section ⓑ may be found from

$$q_b = -h(T_m - T_0)_b \qquad (8.16)$$

(with heat flux inwards being positive) once q_b and h are determined.

The mean heat flux through the pipe wall between sections ⓐ and ⓑ is equal to the enthalpy change of the air:

$$q = \frac{Q}{A} = \frac{\dot{m}c_p(T_b - T_a)}{A}$$

$$= \frac{(\rho UA_c)c_p(T_b - T_a)}{A}$$

(where A_c is the cross-sectional area and A is the heat transfer area of the pipe wall). Substitution yields

$$q = \frac{0.706 \times 1 \times \pi/4 \times (0.02)^2 \times 1.03 \times (-50)}{\pi \times 0.02 \times 1}$$

$$= -0.182 \text{ kW/m}^2$$

and, as q is given as constant along the length of the pipe,

$$q_a = q_b = q$$

The heat transfer coefficient may be found from equation (8.15) as

$$h = \text{Nu} \times \frac{k}{d} = 4.364 \times \frac{4.04 \times 10^{-5}}{0.02}$$

$$= 0.00882 \text{ kW/m}^2 \text{ K}$$

and substitution in equation (8.16) yields T_m at section ⓑ:

$$T_m = \frac{-q_b}{h} + T_0$$

$$= \frac{0.182}{0.00882} + 200$$

$$= 221 \text{ °C}$$

At section ⓐ T_m is 271 °C and a more accurate solution could be obtained using property values at the mean fluid temperature over the length of pipe, of about 246 °C.

8.2.4 Solution of the laminar flow energy equation for constant wall temperature

In this case the rate of change of temperature along the tube $\partial T/\partial x$ (required for substitution in the energy equation (8.5)) is more complicated than in the constant heat flow situation. Equation (8.10) may be differentiated as follows:

$$\frac{\partial}{\partial x}\left(\frac{T_0 - T}{T_0 - T_m}\right) = 0$$

i.e.

$$\frac{\partial}{\partial x}\left(\frac{T_0}{T_0 - T_m}\right) = \frac{\partial}{\partial x}\left(\frac{T}{T_0 - T_m}\right)$$

$$\frac{\partial T_0}{\partial x}\left(\frac{1}{T_0 - T_m}\right) - \frac{T_0}{(T_0 - T_m)^2}\frac{\partial(T_0 - T_m)}{\partial x} = \frac{\partial T}{\partial x}\left(\frac{1}{T_0 - T_m}\right)$$

$$- \frac{T}{(T_0 - T_m)^2}\frac{\partial(T_0 - T_m)}{\partial x}$$

$$\frac{\partial T_0}{\partial x} - \frac{\partial T}{\partial x} = \left(\frac{T_0 - T}{T_0 - T_m}\right)\left(\frac{\partial T_0}{\partial x} - \frac{\partial T_m}{\partial x}\right)$$

$$(8.17)$$

When the wall temperature is constant along the tube, $\partial T_0/\partial x = 0$ and

$$\frac{\partial T}{\partial x} = \left(\frac{T_0 - T}{T_0 - T_m}\right)\frac{\partial T_m}{\partial x}$$

Substitution of this equation and also the velocity profile as given by equation (8.7) into the energy equation (8.5) yields

$$\frac{1}{r}\frac{\partial}{\partial r}\left(r\frac{\partial T}{\partial r}\right) = \frac{2U}{\alpha}\left(1 - \frac{r^2}{r_0^2}\right)\left(\frac{T_0 - T}{T_0 - T_m}\right)\frac{\partial T_m}{\partial x} \qquad (8.18)$$

It is not proposed to test the perseverance of the reader by attempting a solution to this equation in the text. Basically the method is similar to that used in the previous case except that successive approximation is involved in order to evaluate the temperature profile. The final result may be expressed as

$$\text{Nu} = 3.658 \qquad (8.19)$$

The heat transfer coefficient is therefore less in this case than in the case of constant heat flow, and this is due to the slightly different shapes of the temperature profiles.

8.2.5 Empirical considerations regarding laminar flow in tubes

The preceding analysis leads to fairly accurate prediction of the heat transfer rate in ducts for situations in which the laminar flow is fully developed and the properties are constant. In many engineering situations these conditions are not fulfilled and empirical modifications of the basic relationships are used.

The length of tube L required for the attainment of a fully developed velocity profile depends on the tube diameter and the fluid properties. Langhaar (1942) has suggested the following formula for predicting the length of the starting section in the case of laminar flow:

$$\frac{L}{d} = 0.0575 \, \text{Re} \qquad (8.20)$$

With a critical Reynolds number for laminar flow of 2300 the starting length for a 25 mm diameter pipe may therefore be over 3 m. Laminar flow is often encountered in the flow of oils owing to their high viscosity and correspondingly low Reynolds numbers. The viscosity of oil is very sensitive to temperature and can cause significant deviations from heat transfer predictions based on constant property relationships.

A number of empirical and semi-empirical expressions have been developed to take account of the starting length and the property changes. One of the simpler expressions which has withstood the test of time was proposed by Sieder and Tate (1936):

$$\text{Nu} = 1.86 \left(\text{Re Pr} \frac{d}{l} \right)^{0.33} \left(\frac{\mu}{\mu_0} \right)^{0.14} \qquad (8.21)$$

where the properties (including μ) are evaluated at the bulk fluid temperature and μ_0 is the viscosity at the wall temperature which is assumed to be constant. This expression is obviously not applicable to long tubes (where the length l is greater than the starting length L) as it indicates that the heat transfer coefficient tends to zero as l tends to infinity. It is suggested that the value of the heat transfer coefficient predicted by equation (8.21) is only used when it is greater than that predicted by equation (8.19). ·

8.2.6 Empirical relationships for turbulent flow in tubes

The use of Reynolds analogy for the estimation of heat transfer in turbulent conditions has been described and illustrated in Section 3.2. One of the great advantages of this analogy is that it takes some account of irregularities in the pipe geometry (such as thermocouple pockets, bends, T-junctions and internal fins) because the effect on the heat transfer coefficient is similar to the effect on the pressure drop. In addition the effect of surface roughness on heat transfer may be analysed using the variation of the friction factor with roughness.

The reader has already been introduced to the most commonly used empirical heat transfer relationship for smooth pipes, often called the Dittus–Boelter (1930) equation (3.11):

$$\text{Nu} = 0.023 \, \text{Re}^{0.8} \, \text{Pr}^{n} \qquad (8.22)$$

where n is 0.4 for heating fluid and 0.3 for cooling fluid. There are a number of more precise relationships available in the literature for the Nusselt number in turbulent flow in a tube. One of the simplest is

$$Nu = 0.027\ Re^{0.8}\ Pr^{0.33}\left(\frac{\mu}{\mu_0}\right)^{0.14} \tag{8.23}$$

where properties at the bulk temperature are used except for μ_0 which is at the wall temperature. This expression is applicable to fluids with Prandtl numbers in the range 0.7 to 16,700 (Knudsen and Katz, 1958) and over a wide range of temperatures. It requires no correction for the slightly different temperature profiles caused by heating and cooling flow (as does equation (8.22)) because the ratio between the viscosity at the bulk temperature and at the wall compensates for this difference. Although this expression is more widely applicable, the Dittus–Boelter equation is more commonly used as it only requires a knowledge of the bulk temperature of the flow. As in the case of laminar flow, a considerable length of tube may be required for the attainment of the fully developed turbulent flow velocity profile. Generalizations regarding this starting length are not possible owing to sensitivity to the eddy motion at entry, the starting-edge profile and the surface roughness of the tube. For smooth tubes values of L/d in the range 20 to 40 are typical.

Heat transfer to fluid flowing in non-circular ducts is conventionally analysed using the mean effective (or hydraulic) diameter which is defined to yield the same cross-sectional area to perimeter ratio as a circular tube. For a circular tube

$$\frac{\text{Cross-sectional area}}{\text{Perimeter}} = \frac{A_c}{P} = \frac{\pi d^2}{4\pi d} = \frac{d}{4}$$

and the mean effective diameter d_e is defined as

$$d_e = \frac{4A_c}{P} \tag{8.24}$$

For a square of side a, therefore, $d_e = a$ and for an equilateral triangle $d_e = 0.577a$. For an annulus with diameters d_1 and d_2 (with $d_1 > d_2$) substitution yields

$$d_e = \frac{4\pi/4\ (d_1^2 - d_2)^2}{\pi(d_1 + d_2)}$$

$$= (d_1 - d_2)$$

More precisely, P is defined as the 'wetted' perimeter because when the flow of a liquid does not fill the duct section (as in channel flow) only that perimeter over which the fluid flows is included. In the case of turbulent flow the heat transfer expressions developed for circular tubes may be used for non-circular tubes or ducts with the substitution of d_e for d, except when the section is thin enough to cause boundary layer interference.

Example 8.2. In the initial design study of a particular power plant calculations are conducted to determine the effect of various methods of heating the boiler feed water on the overall capital cost and efficiency of the plant. One method involves making use of the geometry of the furnace to incorporate a large coiled duct around the perimeter, and pumping the feed water through this duct before it enters the steam raising part of the furnace.

The duct is of rectangular section measuring 8 cm × 4 cm and the wall temperature is approximately 170 °C throughout. The feed water flows at a rate of 300 kg/min, enters at a temperature of 20 °C and is heated to 150 °C. Compare the heat transfer coefficients obtained using equations (8.22) and (8.23) and estimate the required length of the duct.

Solution. In pipe flow problems involving a considerable fluid temperature change along the length of the pipe it is usual to use property values selected by using the log mean temperature difference. From equation (5.10)

$$\Delta T_m = \frac{150 - 20}{\ln(150/20)} = 64.52\,°C$$

and the property values of water were therefore selected at a temperature of $170 - 64.52 \approx 105\,°C$.

The mean effective diameter d_e of the duct section is given by equation (8.24):

$$d_e = \frac{4A_c}{P} = \frac{4 \times 8 \times 4}{2(8 + 4)} = 5.333\text{ cm}$$

The heat transfer coefficients may now be evaluated from equations (8.22) and (8.23) using property values from the tables of Mayhew and Rogers (1969):

$$\text{Re} = \frac{\rho U d}{\mu} = \frac{\dot{m}}{A}\frac{d}{\mu}$$

$$= \frac{300}{60} \times \frac{5.333 \times 10^{-2}}{32 \times 10^{-4} \times 265 \times 10^{-6}}$$

$$= 3.144 \times 10^5$$

$$\text{Pr} = \frac{c_p \mu}{k} \text{ (at 105 °C)}$$

$$= 1.64$$

Equation (8.22):

$$\text{Nu} = 0.023\,\text{Re}^{0.8}\,\text{Pr}^{0.4}$$

$$= 0.023 \times 2.5 \times 10^4 \times 1.219$$

$$= 700.8$$

$$h = \frac{\text{Nu } k}{d}$$

$$= \frac{700.8 \times 683 \times 10^{-6}}{5.333 \times 10^{-2}}$$

$$= 8.975 \text{ kW/m}^2 \text{ K}$$

Equation (8.23):

$$\text{Nu} = 0.027 \text{ Re}^{0.8} \text{ Pr}^{0.333} \left(\frac{\mu}{\mu_0}\right)^{0.14} \quad (\mu_0 \text{ at } 170 \,^\circ\text{C})$$

$$= 0.027 \times 2.5 \times 10^4 \times 1.179 \left(\frac{265 \times 10^{-6}}{158 \times 10^{-6}}\right)^{0.14}$$

$$= 855.7$$

and $\qquad h = 10.96 \text{ kW/m}^2 \text{ K}$

The latter is probably more accurate as it takes account of viscosity effects. The difference of 18 % in these estimations is typical of the order of accuracy to be expected from empirical convection heat transfer formulae and emphasizes the superfluousness of great precision in determining the property values and temperature differences embodied in these relationships. The duct length *l* may be calculated (using the second value of *h*) as follows:

$$Q = \dot{m} c_p \, \Delta T_{\text{water}}$$

$$= 5 \times 4.226 \times (150 - 20)$$

$$= 2747 \text{ kW}$$

$$Q = hA \, \Delta T_{\text{m}}$$

$$2747 = 10.96 \times 2(0.04 + 0.08)l \times 64.52$$

$$l = 16.19 \text{ m}$$

8.3 Flow Over External Surfaces

8.3.1 The external boundary layer

The external boundary layer differs from the boundary layer within a tube because there is no physical restriction on its thickness. The fully developed velocity and temperature distributions in tube flow increase from zero at the wall to a maximum at the tube centre under all conditions of flow. In the case of a flat plate, which we shall consider in this section, there is no axial symmetry or constant boundary layer thickness and the analysis is inevitably more complicated. For this reason we shall restrict our theoretical analysis to an approximate solution for laminar flow over a flat plate using the heat flow (or integrated boundary layer energy) equation given in Section 8.1.

The equations required in this treatment are generally derived in the text, but the reader is requested in the interest of brevity to accept the common equation for the boundary layer thickness over a flat plate as derived in most fluid dynamics books:

$$\frac{\delta}{x} = \frac{4.64}{\text{Re}_x^{0.5}} \tag{8.25}$$

where δ is the boundary layer thickness and Re_x is the local Reynolds number at a distance x from the leading edge. There is some ambiguity in the term 'boundary layer thickness' as the velocity and temperature distributions from the plate approach the free-stream values asymptotically. In more rigid studies differentiation is made between a boundary layer defined in terms of the displacement of the free stream (displacement thickness) and a boundary layer defined in terms of the drag on the plate (momentum thickness). Furthermore, equation (8.25) is based on a classical but approximate solution to the laminar boundary layer using the momentum integral equation and is therefore not exact. Further information on these and other basic aspects of the boundary layer may be obtained from Kays (1966) and Schlichting (1960). For our purposes we shall content ourselves with the following analysis, which leads to useful results and shall consider some common empirical relationships relating to flat plates and flow around tubes.

8.3.2 The velocity and temperature profiles for laminar flow over a flat plate

The momentum equation of the boundary layer, equation (8.2), is

$$u\frac{\partial u}{\partial x} + v\frac{\partial u}{\partial y} = v\frac{\partial^2 u}{\partial y^2}$$

and at the wall (where u and v are zero)

(i) $\dfrac{\partial^2 u}{\partial y^2} = 0$ at $y = 0$

Referring to Fig. 8.4 the following boundary conditions are applicable:

  (ii) $u = u_\infty$ at $y = \delta$

  (iii) $u = 0$ at $y = 0$

  (iv) $\dfrac{\partial u}{\partial y} = 0$ at $y = \delta$

It is assumed that the form of the velocity profile is constant and may be expressed by the first few terms of a polynomial in y:

$$u = A + By + Cy^2 + Dy^3 + \ldots$$

163

therefore

$$\frac{\partial u}{\partial y} = B + 2Cy + 3Dy^2$$

$$\frac{\partial^2 u}{\partial y^2} = 2C + 6Dy$$

Substitution of boundary conditions (i) to (iv) yields

$$A = 0; \quad B = \frac{3u_\infty}{2\delta}; \quad C = 0; \quad D = -\frac{u_\infty}{2\delta^3}.$$

and the velocity distribution may be expressed as

$$\frac{u}{u_\infty} = \frac{3}{2}\left(\frac{y}{\delta}\right) - \frac{1}{2}\left(\frac{y}{\delta}\right)^3 \tag{8.26}$$

The energy equation of the boundary layer, equation (8.3), is

$$u\frac{\partial T}{\partial x} + v\frac{\partial T}{\partial y} = \alpha\frac{\partial^2 T}{\partial y^2}$$

and rearrangement so that the temperature varies from zero at the wall to a maximum value in the free stream gives

$$u\frac{\partial \Delta T}{\partial x} + v\frac{\partial \Delta T}{\partial y} = \alpha\frac{\partial^2 \Delta T}{\partial y^2}$$

where $\Delta T = T - T_0$. The maximum difference in the free stream is denoted by $\Delta T_\infty = T_\infty - T_0$. This equation is similar to equation (8.2) and evaluation using similar boundary conditions yields

$$\frac{\Delta T}{\Delta T_\infty} = \frac{3}{2}\left(\frac{y}{\delta_t}\right) - \frac{1}{2}\left(\frac{y}{\delta_t}\right)^3 \tag{8.27}$$

where δ_t refers to the thickness of the thermal boundary layer (which for our purposes may be considered as the distance from the wall to the limit of the thermal influence of the wall).

8.3.3 Solution of the heat flow equation for laminar flow over a flat plate at constant temperature

The heat flow equation (8.4) has been presented as

$$\frac{d}{dx}\left(\int_0^Y (T_\infty - T)u\, dy\right) = \alpha\left(\frac{dT}{dy}\right)_0$$

and substitution of $(T_\infty - T) = (\Delta T_\infty - \Delta T)$ from equation (8.27) and u from equation (8.26) yields

$$\frac{d}{dx}\left[\Delta T_\infty u_\infty \int_0^Y \left\{1 - \frac{3}{2}\left(\frac{y}{\delta_t}\right) + \frac{1}{2}\left(\frac{y}{\delta_t}\right)^3\right\}\left\{\frac{3}{2}\left(\frac{y}{\delta}\right) - \frac{1}{2}\left(\frac{y}{\delta}\right)^3\right\} dy\right] = \alpha\left(\frac{d\,\Delta T}{dy}\right)_0$$

It is assumed that δ_t is less than δ and, since we are only concerned with the boundary layer, Y may be replaced by δ. Evaluation of the integral and substitution for $(d\,\Delta T/dy)_0$ from the differential of equation (8.27) leads to

$$\frac{d}{dx}\left[3\,\Delta T_\infty u_\infty\left(\frac{\delta_t^2}{20\delta} - \frac{\delta_t^4}{280\delta^3}\right)\right] = \frac{3}{2}\frac{\alpha\,\Delta T_\infty}{\delta_t}$$

Consideration of the orders of magnitude involved in the two terms in brackets indicates that the second term is negligible in comparison to the other term, yielding

$$u_\infty \frac{d}{dx}\left(\frac{\delta_t^2}{\delta}\right) = \frac{10\alpha}{\delta_t}$$

The classical solution to this equation involves expressing the ratio δ_t/δ as ζ giving

$$u_\infty \frac{d}{dx}(\delta\zeta^2) = \frac{10\alpha}{\zeta\delta} \tag{8.28}$$

The equation may now be solved by substituting for δ from equation (8.25) where

$$\delta = \frac{4.64\,x}{\left(\dfrac{u_\infty x}{\nu}\right)^{0.5}} = 4.64\,x^{0.5}\left(\frac{\nu}{u_\infty}\right)^{0.5}$$

giving

$$\frac{d}{dx}(\zeta^2 x^{0.5}) = \frac{10}{(4.64)^2}\frac{\alpha}{\nu}\frac{1}{\zeta x^{0.5}}$$

Differentiating by parts and noting that

$$\zeta^2\frac{d\zeta}{dx} = \frac{1}{3}\frac{d}{dx}(\zeta^3)$$

yields

$$\frac{4x}{3}\frac{d}{dx}(\zeta^3) + \zeta^3 = 0.9286\left(\frac{\alpha}{\nu}\right)$$

The solution of this first-order linear equation is

$$\zeta^3 = Cx^{-3/4} + 0.9286\,\frac{\alpha}{\nu}$$

With the thermal boundary layer commencing at a distance x_0 from the leading edge, the boundary condition $\zeta = 0$ at $x = x_0$ may be applied to yield

165

$$\zeta^3 = 0.9286 \frac{\alpha}{v}\left[1 - \left(\frac{x}{x_0}\right)^{-3/4}\right]$$

For the case when it commences at the leading edge as shown in Fig. 8.4, $x_0 = 0$, and substituting $v/\alpha = \text{Pr}$ gives

$$\zeta = 0.9746\,\text{Pr}^{-1/3} \qquad (8.29)$$

An expression for the heat transfer coefficient may now be evaluated.

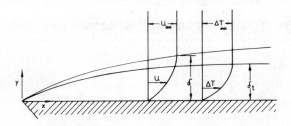

Figure 8.4   Velocity and temperature profile in the boundary layer.

The heat flow rate from a fluid to a flat plate as shown in Fig. 8.4, at a particular distance x from the leading edge, is given by the temperature gradient in the fluid at this point:

$$q_x = -k\left(\frac{\partial T}{\partial y}\right)_{y=0}$$

Also $q_x = -h_x(T_0 - T_\infty) = -h_x\,\Delta T_\infty$ and therefore

$$h_x = \frac{k}{\Delta T_\infty}\left(\frac{\partial T}{\partial y}\right)_{y=0} \qquad (8.30)$$

(It must be emphasized that this equation yields the heat transfer coefficient h_x at a particular distance from the leading edge and is not the mean coefficient which we have so far been using. The suffix x is included to differentiate this local heat transfer coefficient from the mean coefficient.) The differential term may be found from equation (8.27):

$$\left(\frac{\partial T}{\partial y}\right)_{y=0} = \left(\frac{\partial \Delta T}{\partial y}\right)_{y=0} = \frac{3}{2}\left(\frac{\Delta T_\infty}{\delta_t}\right) = \frac{3}{2}\left(\frac{\Delta T_\infty}{\zeta\delta}\right)$$

Substitution for ζ from equation (8.29) and δ from equation (8.25) yields

$$\left(\frac{\partial T}{\partial y}\right)_{y=0} = \frac{3}{2}\left(\frac{\Delta T_\infty}{0.9746\,\text{Pr}^{-1/3}}\frac{\text{Re}_x^{1/2}}{4.64\,x}\right)$$

$$= 0.3317\frac{\Delta T_\infty}{x}\,\text{Re}_x^{1/2}\,\text{Pr}^{1/3}$$

Substitution into equation (8.30) gives

$$h_x = 0.332 \frac{k}{x} \text{Re}_x^{1/2} \text{Pr}^{1/3} \tag{8.31}$$

Generally the mean heat transfer coefficient h is required in heat transfer calculations and this may be found by integrating h_x over the length d of the plate:

$$h = \frac{\int_0^d h_x \, dx}{\int_0^d dx}$$

Substituting for h_x (and noting that Re is a function of x)

$$h = \frac{\int_0^d 0.332 \, k \, \text{Pr}^{1/3} \left(\frac{\rho u_\infty}{\mu}\right)^{1/2} \left(\frac{1}{x}\right)^{1/2} dx}{\int_0^d dx}$$

i.e.

$$h = 2h_x \quad (\text{at } x = d) \tag{8.32}$$

and also

$$\text{Nu} = 2 \, \text{Nu}_x$$

Substituting for h_x in equation (8.31) (with mean Re and Nu based on d) yields

$$\text{Nu} = 0.664 \, \text{Re}^{1/2} \, \text{Pr}^{1/3} \tag{8.33}$$

This is the commonly used equation for heat transfer from a flat plate of constant temperature to laminar flow and is included in Table 3.2. A more exact analysis taking account of variation of fluid properties yields the same relationship. The fluid properties used in this relationship are usually taken at a fluid temperature T_f which is the arithmetic mean of the wall and free-stream temperatures:

$$T_f = \frac{T_0 + T_\infty}{2}$$

Example 8.3. Heavy grade oil is passed through a heat exchanger to slightly increase its temperature so that the viscosity and subsequently the pumping power is reduced. The exchanger consists of 20 hollow flat plates situated in the oil stream as shown in Fig. 8.5. Hot water flows through the

167

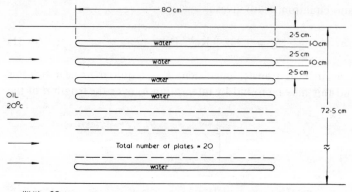

Width = 80 cm

Figure 8.5 Heat exchanger arrangement.

plates (in the direction normal to the plane of the figure) at such a rate that the surfaces of the plates are at an approximately uniform temperature of 100 °C.

By making reasonable assumptions estimate the local heat transfer coefficient a third of the way along the plates, the mean heat transfer coefficient and the temperature rise of the oil. Oil enters the exchanger at 20 °C and the flow rate between the plates is 1 m/s. The properties of the oil at the mean fluid temperature of about 60 °C may be taken as:

$$\rho = 850 \text{ kg/m}^3 \qquad\qquad k = 0.18 \times 10^{-3} \text{ kW/m K}$$
$$\mu = 8 \times 10^{-3} \text{ kg/m s} \qquad\qquad c_p = 1.1 \text{ kJ/kg K}$$

Solution. For a flat plate of length 0.8 m:

$$\text{Re} = \frac{\rho u_\infty d}{\mu} = \frac{850 \times 1 \times 0.8}{8 \times 10^{-3}}$$
$$= 85,000$$

Since this is less than the critical Reynolds number for transition to turbulent flow on a flat plate of 500,000 (see Section 3.3); the flow is laminar. If flat plate laminar flow theory is to be applied, it must be checked that the boundary layers between the plates do not meet within the length of the plates. From equation (8.25)

$$\delta = \frac{4.64 \, x}{\text{Re}_x^{0.5}}$$

$$= \frac{4.64 \times 0.8}{(85,000)^{0.5}} = 0.0127 \text{ m}$$

and two boundary layers (each of 1.27 cm thick) will therefore meet within the plate spacing of 2.5 cm. Nevertheless, it is reasonable to apply the flat plate theory at least as a first estimate because the layers converge gradually and will only meet in the last part of the plate length. It is also assumed in this

168

estimation that the effects of natural convection, the rounded leading edge and property variation with temperature are of a secondary nature. From equation (8.31) the local Nusselt number Nu_x is

$$Nu_x = \frac{h_x x}{k} = 0.332 \, Re_x^{1/2} \, Pr^{1/3}$$

and substitution at $x = 0.8/3$ yields

$$\frac{h_x \times 0.267}{0.18 \times 10^{-3}} = 0.332 \left(\frac{850 \times 1 \times 0.267}{8 \times 10^{-3}}\right)^{1/2} \left(\frac{1.1 \times 8 \times 10^{-3}}{0.18 \times 10^{-3}}\right)^{1/3}$$

from which

$$h_x = 0.205 \text{ kW/m}^2 \text{ K at } x = 0.267 \text{ m}$$

From equation (8.33) the mean heat transfer coefficient is

$$Nu = \frac{hd}{k} = 0.664 \, Re^{1/2} \, Pr^{1/3}$$

and on substitution

$$h = 0.159 \text{ kW/m}^2 \text{ K}$$

The overall heat transfer in the exchanger is given by

$$Q = hA \, \Delta T_m = (\dot{m} c_p \, \Delta T)_{oil}$$

where A is the total heat exchange area of $40 \times 0.8 \times 0.8 = 25.6 \text{ m}^2$. The term ΔT_m cannot be determined directly but, since it is indicated that the oil temperature rise is slight, a reasonable approximation is that $\Delta T_m = 80\,°C$. The mass flow rate of the oil is given by

$$\dot{m} = \rho U A_c$$

where A_c is the cross-sectional flow area of $(0.025 \times 0.8) \times 21 \text{ (gaps)} = 0.42 \text{ m}^2$ Thus substitution yields

$$0.159 \times 25.6 \times 80 = 850 \times 1 \times 0.42 \times 1.1 \times \Delta T_{oil}$$

and

$$\Delta T_{oil} = 0.83\,°C$$

With the assumptions and simplifications made in this estimation the result can only be taken as indicating that the oil temperature increases by around $1\,°C$. Although this increase is slight, it can alter the viscosity of some oils by 10% and reduce the pumping power required to transport the oil by a similar amount.

8.3.4 Turbulent flow over a flat plate

A simple approximate equation for heat transfer under conditions of turbulent flow over a flat plate may be derived from Reynolds analogy, expressed in the form

$$Nu_x = \frac{f}{2} Re_x Pr$$

From fluid dynamics practice the friction factor in turbulent flow is

$$f = \frac{0.059}{Re_x^{0.2}} \tag{8.34}$$

and substitution yields

$$Nu_x = 0.0295\, Re_x^{0.8}\, Pr$$

where suffix x is added to indicate that this applies to the local Nusselt number. The mean Nusselt number for a plate of length d, assuming turbulent flow from the leading edge for the moment, is found in a similar way to the mean value for laminar flow (see equation (8.32)). From the above equation

$$\frac{h_x x}{k} = 0.0295\left(\frac{\rho u_\infty x}{\mu}\right)^{0.8} Pr$$

$$h = \frac{\displaystyle\int_0^d h_x\, dx}{\displaystyle\int_0^d dx}$$

$$= \frac{h_x}{0.8}$$

and

$$Nu = \frac{Nu_x}{0.8}$$

Therefore the heat transfer relationship based on the mean Nusselt number becomes

$$Nu = 0.037\, Re_x^{0.8}\, Pr$$

This gives reasonably accurate results for gases where Pr is about unity and compares favourably with empirical relationships such as the one recommended by McAdams (1954) and originally given by Colburn (1933), which on rearrangement gives

$$Nu = 0.036\, Re^{0.8}\, Pr^{0.33} \tag{8.35}$$

In situations where the fluid transforms from laminar to turbulent flow within the length of plate under study, the overall heat transfer must be evaluated separately in each section. Let the length from the leading edge of the plate to the critical point where turbulent flow commences be x_{crit} and the

Reynolds number at this point be Re_{crit}. If the flow from the leading edge was turbulent, the mean Nusselt number (based on mean h) for the heat transfer from the leading edge to x_{crit} would be predicted by equation (8.35) with Re replaced by Re_{crit}. As the flow is actually laminar in this region a more correct prediction is given by the laminar flow relationship (equation (8.33)):

$$Nu = 0.664 \, Re_{crit}^{0.5} \, Pr^{0.33}$$

and the reduction in the turbulent flow Nusselt number is therefore $(0.036 \, Re_{crit}^{0.8} - 0.664 \, Re_{crit}^{0.5})Pr^{0.33}$. If equation (8.35) is applied to the complete plate, where $x > x_{crit}$ and the reduced Nusselt number in the laminar region is included, it then becomes

$$Nu = 0.036 \, Re^{0.8} \, Pr^{0.33} - (0.036 \, Re_{crit}^{0.8} - 0.664 \, Re_{crit}^{0.5})Pr^{0.33}$$

It was indicated in Section 3.3 that for a smooth plate $Re_{crit} \approx 500{,}000$ and substitution yields, for the mean Nusselt number,

$$Nu = (0.036 \, Re^{0.8} - 836)Pr^{0.33} \qquad (8.36)$$

(where properties are evaluated at the mean fluid temperature T_f).

A less empirical approach to the problem of heat transfer from a flat plate is provided by a sophistication of Reynolds analogy termed the Prandtl–Taylor modification. This modification takes into account the effect of the laminar sub-layer adjacent to the surface within the boundary layer as shown in Fig. 8.6. It is assumed that heat transfer is only possible by conduction in

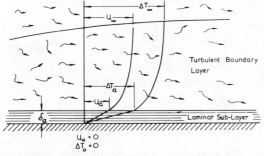

Figure 8.6 The Prandtl–Taylor modification.

the sub-layer of thickness δ_a. Using the nomenclature indicated in the figure the heat conduction through this sub-layer is given by

$$q = \frac{-k}{\delta_a} \Delta T_a$$

and the shear stress at the wall is

$$\tau_0 = \rho v \left(\frac{du}{dy}\right)_0$$

$$= \rho v \frac{u_a}{\delta_a}$$

Elimination of δ_a gives

$$q = -\frac{\tau_0 k}{\rho v u_a} \Delta T_a$$

Application of Reynolds analogy, equation (3.5), to the edge of the sub-layer (noting that $\tau_a = \tau_0$) gives

$$q = -c_p \frac{(\Delta T_\infty - \Delta T_a)}{u_\infty - u_a} \tau_0$$

where it is assumed that the sub-layer is sufficiently thin for the velocity distribution to be linear and consequently the shear stress to be constant throughout the sub-layer.

Elimination of ΔT_a and rearrangement then yields

$$q = \frac{-c_p \Delta T_\infty \tau_0}{u_\infty} \left[\frac{1}{1 + u_a/u_\infty(\mathrm{Pr} - 1)} \right] \tag{8.37}$$

and comparison with equation (3.5) shows that (with $U = u_\infty$) the basic analogy is modified by the term in brackets. From classical fluid dynamic theory the velocity ratio between the sub-layer and the free stream is given by

$$\frac{u_a}{u_\infty} = \frac{2.11}{\mathrm{Re}^{0.1}}$$

and τ_0 is given by equation (3.7). Substitution and algebraic rearrangement into dimensionless form yields for the local Nusselt number

$$\mathrm{Nu}_x = \frac{f}{2} \mathrm{Re}\, \mathrm{Pr}\!\left(\frac{1}{1 + 2.11\,\mathrm{Re}^{-0.1}(\mathrm{Pr} - 1)} \right) \tag{8.38}$$

Combination with equation (8.34) and substitution of $\mathrm{Nu} \approx \mathrm{Nu}_x/0.8$ (approximate in this case as the bracket is slightly dependent on x) then yields

$$\mathrm{Nu} = 0.037\,\mathrm{Re}^{0.8}\, \mathrm{Pr}\!\left(\frac{1}{1 + 2.11\,\mathrm{Re}^{-0.1}(\mathrm{Pr} - 1)} \right) \tag{8.39}$$

Von Karman (1939) extended the analysis to the case of three zones by including a buffer layer next to the laminar sub-layer. Equations (8.34) and (8.35) are less cumbersome in use than the above relationships and are recommended for general engineering purposes.

Example 8.4. Water flows at the rate of 3 m/s over a smooth plate which is 2 m long. The bulk water temperature is 20 °C and the plate temperature is uniform at 80 °C. Determine the heat transfer coefficient at the plate as predicted by the momentum analogy, the Prandtl–Taylor modification and the Colburn empirical relationship, and comment on the results.

Solution. In the absence of further information the relevant properties of water may be taken at the mean temperature of 50 °C from tables (such as those of Mayhew and Rogers, 1969) as

$$\rho = 988.1 \text{ kg/m}^3$$
$$\mu = 0.544 \times 10^{-3} \text{ kg/m s}$$
$$c_p = 4.181 \text{ kJ/kg K}$$
$$k = 0.643 \times 10^{-3} \text{ kW/m K}$$

The dimensionless numbers are:

$$\text{Re} = \frac{\rho u_\infty d}{\mu} = 10.90 \times 10^6$$

$$\text{Pr} = \frac{c_p \mu}{k} = 3.45$$

From Reynolds analogy:

$$\text{Nu} = 0.037 \text{ Re}^{0.8} \text{ Pr}$$
$$= 0.037 \times 426,500 \times 3.54$$
$$= 55,860$$

and

$$h = \frac{\text{Nu } k}{d}$$

$$= \frac{55,860 \times 0.643 \times 10^{-3}}{2}$$

$$\underline{h = 17.96 \text{ kW/m}^2 \text{ K}}$$

From the Prandtl–Taylor modification:

$$\text{Nu} = 0.037 \text{ Re}^{0.8} \text{ Pr} \left[\frac{1}{1 + 2.11 \text{ Re}^{-0.1}(\text{Pr} - 1)} \right]$$

$$= 55,860 \left[\frac{1}{1 + 2.11 \times 0.274 \, (2.54)} \right]$$

$$= 22,630$$

and similarly to the above this leads to

$$\underline{h = 7.28 \text{ kW/m}^2 \text{ K}}$$

From equation (8.35):

$$\text{Nu} = 0.036 \text{ Re}^{0.8} \text{ Pr}^{0.33}$$
$$= 0.036 \times 426,500 \times 1.524$$
$$= 23,400$$

and

$$\underline{h = 7.52 \text{ kW/m}^2 \text{ K}}$$

173

From equation (8.36) which takes account of the initial laminar section:

$$\text{Nu} = 0.036 \, \text{Re}^{0.8} \, \text{Pr}^{0.33} - 836 \, \text{Pr}^{0.33}$$
$$= 23,400 - 1,274$$
$$\approx 22,100$$

and $\qquad h = 7.11 \text{ kW/m}^2 \text{ K}$

Comparing the values of h the following comments may be made.

The value of h obtained using Reynold analogy is obviously high as the basic analogy is only applicable to fluids with a Prandtl number of approximately unity. The Prandtl–Taylor modification leads to a value of h which is rather lower than that given by equation (8.35). Prandtl (1928), himself reported that the constant 2.11 in the formula appeared from experiments to be rather high. Further analysis and experiments showed that this constant is dependent on the Prandtl number (see Eckert and Drake, 1972) and is more correctly replaced by $1.48 \, \text{Pr}^{-0.167}$. In this case the constant is $1.48 \times 3.54^{-0.167} = 1.20$ and this yields $h = 9.78 \text{ kW/m}^2 \text{ K}$ which, if correlated by experiment, would indicate that the value yielded by equation (8.35) is 23% low. Finally, the reduction due to an initial laminar section is about 5% in this case. (If Re_{crit} is taken as 0.5×10^6, x_{crit} is found to be 92 mm.) For values of Reynolds numbers greater than 10^6 it is therefore reasonable to ignore the effect of an initial laminar section as this is well within the likely prediction error.

8.3.5 Cross-flow over a cylinder

A problem commonly encountered in heat exchangers and in general engineering involves the estimation of heat transfer to a cylinder from a fluid flowing in cross-flow (that is normal to the cylinder axis). In this situation a separation of the boundary layer occurs as shown for the case of laminar flow in Fig. 8.7 and the theoretical analysis is prevented by the random

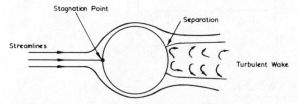

Figure 8.7 Flow over a cylinder.

nature of the swirls and vortices in the wake of the cylinder. In this case therefore we shall simply present the commonly used empirical results for forced convection heat transfer in separated flow for simple geometries. Experimental correlations for heat transfer in cross-flow to banks of tubes in various geometrical arrangements and to tubes with extended surfaces are available in reference works such as Kern (1950), McAdams (1954) and Jakob (1949, 1957).

The empirical data are usually presented in the form

$$Nu = C\ Re^n \tag{8.40}$$

and values of C and n for flow over a cylinder under various conditions are presented in Table 8.1. These values are taken from Jakob (1949, 1957) where more extensive data are available. In the case of non-circular rods the

TABLE 8.1 Heat transfer correlation for gas in cross-flow over a cylinder (Hilpert, 1933)

$Nu = C\ Re^n$		
Re	C	n
1–4	0.891	0.330
4–40	0.821	0.385
40–4000	0.615	0.466
4000–40,000	0.174	0.618
40,000–250,000	0.0239	0.805

physical dimension d of the Nusselt and Reynolds numbers is taken as the diameter of a rod which would have the same surface area, (i.e., not the mean hydraulic diameter). Equation (8.40) with the constants of Table 8.1 is strictly only applicable to gases and when applied to liquids yields values which are generally rather less than those encountered in practice. Knudsen and Katz (1958) present results for liquid flow normal to a cylinder which effectively modify equation (8.40) to

$$Nu = 1.11\ C\ Re^n\ Pr^{1/3} \text{ (for liquids)} \tag{8.41}$$

where C and n are given in Table 8.1 as before. McAdams (1954) suggests the relationship

$$Nu = 0.37\ Re^{0.6} \tag{8.42}$$

for the heat transfer from a flowing gas to a sphere in the Reynolds number range 17 to 70,000. In each of these expressions (8.40, 8.41 and 8.42) the properties are evaluated at the film temperature T_f defined as the arithmetic mean of the body surface and free-stream temperatures.

8.4 Derivation of the Boundary Layer Equations

The simplifications of the general viscous fluid flow theory which may be applied in the case of a boundary layer were enumerated in Section 8.1. In

175

this section these simplifications are assumed to be acceptable and are not repeated. Equations (8.1) to (8.4) may be derived as follows.

A control volume of unit depth in the z-direction positioned within a laminar boundary layer is shown in Fig. 8.1. In Fig. 8.8 the fluid mass flow

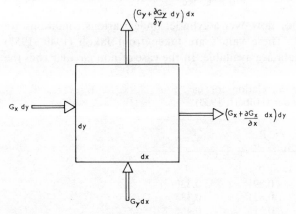

Figure 8.8   Mass flow through the control volume.

rates per unit area ($G = \dot{m}/A_c = \rho U$) across the control volume boundaries in conventional Cartesian coordinates are indicated. Assuming there is no mass storage, summation of the differences between ingoing and outgoing flows in each direction yields

$$\frac{\partial G_x}{\partial x} + \frac{\partial G_y}{\partial y} = 0 \tag{8.43}$$

and as $G_x = \rho u$, $G_y = \rho v$ and the density is constant

$$\frac{\partial u}{\partial x} + \frac{\partial v}{\partial y} = 0 \tag{8.1}$$

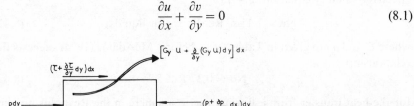

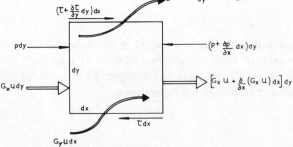

Figure 8.9   Force balance on the control volume.

Equating the changes of momentum flux and external forces on the control volume of unit depth as shown in Fig. 8.9 gives

$$\frac{\partial}{\partial x}(G_x u) + \frac{\partial}{\partial y}(G_y u) = \frac{\partial \tau}{\partial y} - \frac{\partial p}{\partial x}$$

Expanding the left-hand terms and equating τ to $\mu\, \partial u/\partial y$ yields

$$G_x \frac{\partial u}{\partial x} + \frac{u\, \partial G_x}{\partial x} + G_y \frac{\partial u}{\partial y} + \frac{u\, \partial G_y}{\partial y} = \frac{\partial}{\partial y}\left(\mu \frac{\partial u}{\partial y}\right) - \frac{\partial p}{\partial x}$$

Making use of equation (8.43), assuming μ is constant and neglecting the pressure term as mentioned earlier:

$$G_x \frac{\partial u}{\partial x} + G_y \frac{\partial u}{\partial y} = \mu \frac{\partial^2 u}{\partial y^2}$$

Substitution of $G_x = \rho u$, $G_y = \rho v$ and the kinematic viscosity $v = \mu/\rho$ then gives

$$u \frac{\partial u}{\partial x} + v \frac{\partial u}{\partial y} = v \frac{\partial^2 u}{\partial y^2} \tag{8.2}$$

More complex analysis of the general case in three dimensions including the effects of changes with time t and external body forces per unit mass of fluid X, Y and Z, yields the Navier–Stokes equations for constant-property fluids:

$$u \frac{\partial u}{\partial x} + v \frac{\partial u}{\partial y} + w \frac{\partial u}{\partial z} + \frac{\partial u}{\partial t} = v\left(\frac{\partial^2 u}{\partial x^2} + \frac{\partial^2 u}{\partial y^2} + \frac{\partial^2 u}{\partial z^2}\right) - \frac{1}{\rho}\frac{\partial p}{\partial x} + X$$

$$u \frac{\partial v}{\partial x} + v \frac{\partial v}{\partial y} + w \frac{\partial v}{\partial z} + \frac{\partial v}{\partial t} = v\left(\frac{\partial^2 v}{\partial x^2} + \frac{\partial^2 v}{\partial y^2} + \frac{\partial^2 v}{\partial z^2}\right) - \frac{1}{\rho}\frac{\partial p}{\partial y} + Y$$

$$u \frac{\partial w}{\partial x} + v \frac{\partial w}{\partial y} + w \frac{\partial w}{\partial z} + \frac{\partial w}{\partial t} = v\left(\frac{\partial^2 w}{\partial x^2} + \frac{\partial^2 w}{\partial y^2} + \frac{\partial^2 w}{\partial z^2}\right) - \frac{1}{\rho}\frac{\partial p}{\partial z} + Z$$

where u, v and w are the velocities in directions x, y and z. Equation (8.2) is therefore a special case of the Navier–Stokes equations which is particularly useful for the study of thin boundary layers.

The energy transfer across the control volume (of Fig. 8.1) consists essentially of convection in the x- and y-directions and conduction in the y-direction only. Conduction in the x-direction and the viscous energy dissipation is assumed negligible. Summation of the differences between the ingoing and outgoing terms indicated in Fig. 8.10 gives

$$\frac{\partial}{\partial x}(G_x T) + \frac{\partial}{\partial y}(G_y T) = \frac{k}{c_p}\frac{\partial}{\partial y}\left(\frac{\partial T}{\partial y}\right)$$

Expanding the left-hand terms gives

$$G_x \frac{\partial T}{\partial x} + T \frac{\partial G_x}{\partial x} + G_y \frac{\partial T}{\partial x} + T \frac{\partial G_y}{\partial x} = \frac{k}{c_p}\frac{\partial^2 T}{\partial y^2}$$

177

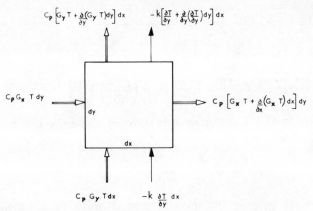

Figure 8.10   Energy transfer through the control volume.

Making use of equation (8.43) and substituting $G_x = \rho u$, $G_y = \rho v$ and $\alpha = k/(\rho c_p)$ yields

$$u\frac{\partial T}{\partial x} + v\frac{\partial T}{\partial y} = \alpha\frac{\partial^2 T}{\partial y^2} \tag{8.3}$$

Analysis of the general case in three dimensions yields, for constant property fluids,

$$u\frac{\partial T}{\partial x} + v\frac{\partial T}{\partial y} + w\frac{\partial T}{\partial z} + \frac{\partial T}{\partial t} = \alpha\left(\frac{\partial^2 T}{\partial x^2} + \frac{\partial^2 T}{\partial y^2} + \frac{\partial^2 T}{\partial z^2}\right) + V$$

where $V = f(\mu, u, v, w)$ and denotes the viscous energy transfer. Equation (8.3) is therefore a special case of the energy equation which is applicable under the boundary layer conditions outlined in Section 8.1. Some authors term this the heat flow equation as it only involves heat flow terms, but here we shall reserve this name for the integral form of the energy equation which may be derived as follows.

In this development of the integral energy equation for the boundary layer, we consider a section of unit depth bounded by the surface and the free stream as indicated in Fig. 8.11. The energy transfer into the vertical side is given by $(\dot{m}_x c_p T)/A = G_x c_p T$ integrated over the side from an arbitrary point in the free stream Y to the wall. The energy transfer into the upper side $(y = Y)$ is the product of the mass flux and $c_p T$. Since the lower edge has zero mass flux the required mass flux is the difference in the mass fluxes through the right-hand and left-hand sides, given by

$$\left[\int_0^Y G_x \, \mathrm{d}y + \frac{\mathrm{d}}{\mathrm{d}x}\left(\int_0^Y G_x \, \mathrm{d}y\right)\mathrm{d}x\right] - \int_0^Y G_x \, \mathrm{d}y$$

The temperature of the fluid at the upper side is constant at T_∞ and the energy transfer may therefore be expressed as

178

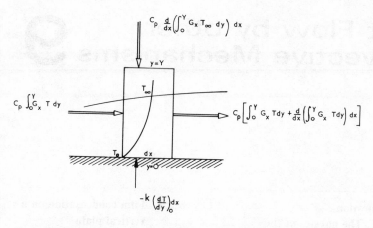

Figure 8.11 Control volume for derivation of the integral energy equation.

$$c_p \frac{d}{dx}\left(\int_0^Y G_x T_\infty \, dy\right) dx$$

At the wall ($y = 0$) there is conduction heat transfer only.

Summation of the ingoing and outgoing energy terms shown in Fig. 8.11 yields the following expression

$$c_p \frac{d}{dx}\left(\int_0^Y G_x T \, dy\right) dx - c_p \frac{d}{dx}\left(\int_0^Y G_x T_\infty \, dy\right) dx = -k\left(\frac{dT}{dy}\right)_0 dx$$

Rearrangement, noting that $G_x = \rho u$ and $\alpha = k/(\rho c_p)$ gives the integral energy equation for the boundary layer under conditions of negligible viscous energy and constant free-stream temperature:

$$\frac{d}{dx}\left(\int_0^Y (T_\infty - T)u \, dy\right) = \alpha\left(\frac{dT}{dy}\right)_0 \qquad (8.4)$$

Derivations of more generalized boundary layer equations are available in texts on boundary layer theory such as Schlichting (1960).

Heat Flow by other Convective Mechanisms 9

9.1 Evaporation

The evaporation of a liquid to a vapour is one of the most familiar and important heat transfer processes. It occurs in the boiler of a steam power plant, the evaporator of a refrigeration cycle and the distillation column of a chemical plant. In spite of its importance, boiling was one of the last heat transfer processes to be studied and it was not until the research of Nukiyama (1934) in Japan that the various regimes of boiling were properly identified. During the 1960s a vast amount of both pure and applied research was conducted on the boiling process mainly as a result of the precise data required for the design of steam generation plant for nuclear reactors. In this section the reader's attention is directed towards evaporation due to boiling rather than evaporation at sub-saturation temperatures as in drying processes. (Evaporation at sub-saturation temperatures is generally considered as an aspect of mass transfer and an introduction to this topic is given in the text by Eckert and Drake, 1972.) Furthermore, the treatment is restricted to nucleate boiling as opposed to other types of boiling not commonly encountered in engineering systems.

9.1.1 The physics of the boiling process

Let us consider the case when heat is added to a stationary quantity of fluid from an immersed heating surface. This type of boiling situation is termed 'pool boiling' to distinguish it from 'convective boiling' where the fluid is flowing over the surface. A typical pool boiling experimental rig with a horizontal heating surface is shown in Fig. 9.1 and consists essentially of a

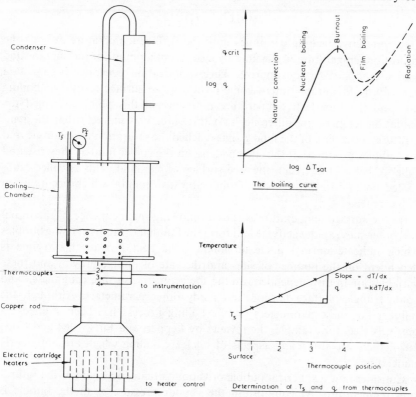

Figure 9.1 Pool boiling rig.

boiling chamber containing a copper heating surface and a condenser arranged so that the condensate is returned to the chamber. The main parameters of interest are the bulk fluid temperature T_f and pressure p_f, the metal surface temperature T_s and the heat flux q at the surface. The boiling heat transfer coefficient h is based on the difference between the saturation temperature T_{sat} of the fluid at p_f and the surface temperature:

$$q = -h(T_{sat} - T_s) = h \Delta T_{sat} \qquad (9.1)$$

where $\Delta T_{sat} = T_s - T_{sat}$. Both T_s and q may be experimentally determined by measuring the temperature at a number of known points along the axis of the copper rod conducting heat to the surface. This is conveniently achieved by inserting thermocouple junctions in small radial holes in the rod. Extrapolation of a temperature–position plot then yields the surface temperature and the slope yields the heat flux as indicated in Fig. 9.1.

We shall now proceed to describe the physical and thermal changes which

181

occur as the surface temperature is increased. Initially as the surface temperature rises above the fluid temperature natural or free convection currents are induced and (as indicated in Section 9.3):

$$q \propto (T_s - T_f)^{5/4}$$

Free convective heating continues until the surface temperature exceeds the saturation temperature of the fluid by a sufficient amount for the superheated fluid at the surface to vaporize. This vaporization generally occurs at a particularly favourable microscopic cavity in the surface termed a 'boiling site' and the superheat temperature difference at this point is termed the 'initiation temperature difference'. At this stage it is unlikely that the temperature of the bulk of the fluid T_f has reached the saturation temperature and the bubbles produced at the site condense in the cooler fluid as they rise and collapse before they reach the free surface of the fluid. This is called 'subcooled boiling' to distinguish it from 'saturated boiling' which occurs when T_f has reached T_{sat}.

As the surface temperature is raised following the incipience of the initial site, other sites rapidly activate and form the familiar columns of rising bubbles which indicate normal nucleate boiling. The heat flux in this regime is dependent on the number of sites and the rate of bubble growth and these variables are in turn dependent on the site geometry, surface roughness, fluid contact angle, surface tension and many other parameters. Although the influence of these parameters on the boiling process has been studied extensively (see, for example, the reviews by Leppert and Pitts, 1964 and Cole, 1974) there are still many aspects of nucleate boiling which are not clearly understood.

Consider now a spherical bubble of vapour situated in liquid at the saturation temperature. At equilibrium the vapour pressure p_g in the bubble is balanced by the liquid pressure p_f and the surface tension σ as indicated in Fig. 9.2a. Resolving the forces vertically:

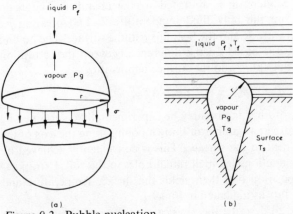

Figure 9.2 Bubble nucleation.

$$\pi r^2 p_g = \pi r^2 p_f + 2\pi r \sigma$$

therefore

$$p_g - p_f = \frac{2\sigma}{r} \tag{9.2}$$

The pressure difference between gaseous and liquid phases at equilibrium is given by a thermodynamic relationship termed the Clausius–Clapeyron equation:

$$\frac{dp}{dT} = \frac{h_{fg}}{v_{fg} T_{sat}}$$

where h_{fg} is the enthalpy of vaporization and v_{fg} is the change of specific volume between the phases. In finite difference form, assuming the liquid specific volume is negligible compared to the vapour specific volume v_g (and noting density $\rho_g = 1/v_g$), this may be written as

$$\frac{p_g - p_f}{T_g - T_{sat}} = \frac{\rho_g h_{fg}}{T_{sat}} \tag{9.3}$$

where T_g refers to the vapour temperature and T_{sat} to the saturation temperature at p_f. Combination of equations (9.2) and (9.3) yields on rearrangement

$$r = \frac{2\sigma T_{sat}}{\rho_g h_{fg}(T_g - T_{sat})} \tag{9.4}$$

Thus the equilibrium bubble size is a function of the excess temperature of the vapour over the saturation temperature of the liquid. Alternatively, since at equilibrium when the bubble is neither expanding nor contracting (that is neither evaporating nor condensing), $T_f = T_g$ it follows that r is a function of the superheat temperature difference $(T_f - T_{sat})$. The bubble therefore only exists in the metastable state of the liquid and as the superheating tends to zero r tends to infinity and the stable interface becomes planar.

A useful idealization for bubble growth at a surface is to consider that a site initiates by vapour growth within a conical cavity and reaches a critical point when a hemispherical bubble sits on the mouth of the cavity, as shown in Fig. 9.2b. It can be shown that further bubble growth is likely to result in the bubble leaving the surface (Bankoff, 1958) and in this situation the critical bubble radius may be approximately equated to the mouth radius r_c of the cavity. If it is assumed that the vapour temperature T_g may be equated to the temperature of the cavity, that is the surface temperature T_s, equation (9.4) yields an expression for the cavity size:

$$r_c = \frac{2\sigma T_{sat}}{\rho_g h_{fg}(T_s - T_{sat})}$$

that is

$$r_c = \frac{2\sigma T_{sat}}{\rho_g h_{fg} \Delta T_{sat}} \tag{9.5}$$

where T_{sat} is in degrees Kelvin.

In the case of water boiling at atmospheric pressure with a typical ΔT_{sat} value of about 10 °C, the value of r_c predicted by this expression is about 3.5 μm (0.0035 mm or 0.14×10^{-3} inches). Studies under isothermal conditions by Griffith and Wallis (1960) and non-isothermal conditions by Shoukri and Judd (1975) and Cornwell (1975) have confirmed that cavities are typically of this size. Thus enhanced nucleate boiling can be achieved by the use of rough or porous surfaces which have an abundance of cavities of the right size. Tubes with a porous heat transfer surface specially produced for boiling applications are now available commercially.

Returning now to our consideration of pool boiling on the copper surface, the number of active sites increases as the surface temperature and heat flux increase until a point is reached where there is interference and coalescence of the bubbles. This gives rise to a point of inflection on the q versus ΔT_{sat} curve and is an indication that the maximum or critical heat flux is being approached. The critical heat flux occurs when the bubbles coalesce to such an extent that they blanket the heater surface with vapour and prevent the flow of liquid to the surface. Under this condition the layer of vapour effectively insulates the surface from the liquid and the surface temperature therefore rises to a high value. Boiling in this manner is termed 'film boiling' and, as the temperature rise is sometimes sufficient to melt the heating surface, the critical point is often termed 'burnout'. If the surface temperature is raised above the value required for stable film boiling (typically a few hundred degrees) radiation across the vapour layer becomes significant and results in a heat flux dependence on T_s^4. The experimental rig of Fig. 9.1 is unsuitable for study of boiling at these high temperature differences and film boiling is generally demonstrated using an electrically-heated submerged wire as the heat source.

In engineering situations burnout is generally avoided by careful thermal design to ensure that the working range is well within the nucleate boiling regime. Nevertheless, notwithstanding the factor of safety required, extremely high heat fluxes and heat transfer coefficients are possible. The heat flux curve for water boiling on a copper surface (finished with 4/0 grade emery paper) is shown in Fig. 9.3. This curve was obtained at one atmosphere; at the higher pressures encountered in steam power plant boilers the curve is situated further to the left of the figure so that the heat flux obtained at a particular ΔT_{sat}, (and also the heat transfer coefficient), is increased. Furthermore, in modern power station boilers the water and steam are heated as they flow through the boiler tubes and forced-convective or flow boiling occurs which leads to even higher heat fluxes. For an introduction to boiling and two-phase flow the reader is referred to the texts by Tong (1965) and Collier (1972).

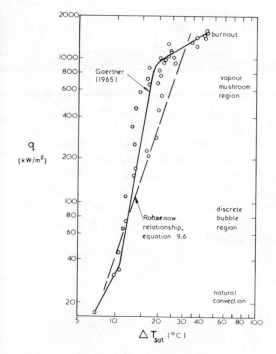

Figure 9.3 Heat transfer data for water boiling on a 4/0 polished copper surface at 1 atmosphere (adapted from Gaertner, 1965).

9.1.2 Empirical boiling heat transfer correlations

The high heat transfer coefficients encountered in nucleate boiling heat transfer are not primarily due to the latent heat transported by the bubbles from the surface. A simple calculation of the latent heat transport taking account of the number, size and frequency of the bubbles at the surface indicates that typically only about 2% of the total heat is transferred in this way. The majority of the heat is transferred to the bubbles as they rise through the liquid by the violent convection currents caused by the stirring action of the bubbles. The intense agitation in the liquid layer adjacent to the surface may be simulated by passing gas through small holes in a surface under normal forced convection conditions; experiments of this nature have yielded similar heat transfer coefficients to those experienced in nucleate boiling. Consequently, empirical nucleate boiling correlations are based on convection and are of a similar general form to convection relationships.

In the boiling situation the definition of the Reynolds number is not obvious and studies by various researchers have led to a number of correlations. For example, Rohsenow (1952) developed a relationship (based on a Reynolds number involving a mean velocity of the vapour from the surface) which is generally presented in the form:

185

$$\frac{c_f \, \Delta T_{sat}}{h_{fg}} = C_b \left[\frac{q}{\mu_f h_{fg}} \left(\frac{\sigma}{g(\rho_f - \rho_g)} \right)^{1/2} \right]^{1/3} Pr_f^{1.7} \tag{9.6}$$

where C_b, the Rohsenow boiling constant, has values for water as shown in Table 9.1 and $\sigma = 59.3 \times 10^{-3}$ N/m for water at 100 °C. While this equation gives reasonable approximate prediction of the q–ΔT relationship more detailed studies (Vachon *et al.* (1968), Mikic and Rohsenow (1969) and others) show that the relationship depends upon the form of nucleate boiling as indicated in Fig. 9.3. In particular the dependence of q on ΔT^3 indicated in equation (9.6) is a mean relationship over the complete nucleate boiling range and experiment shows that the power varies from about 6 to 0.5 from the incipience of boiling to burnout. Another correlation which is sometimes used and leads to satisfactory approximate prediction of the nucleate boiling curve is due to Forster and Zuber (1955) and is based on a Reynolds number involving the velocity of the bubble boundary.

TABLE 9.1 C_b values for boiling water

Surface	C_b
Platinum	0.013
Nickel	0.006
Copper	0.013
Stainless steel	0.014
Brass	0.006

The importance of prediction of the peak nucleate boiling or burnout heat flux q_{crit} has been mentioned. A number of empirical correlations and also theoretical analyses (such as Zuber, 1958) are available in the technical literature. In practice burnout has been found to be sensitive to surface roughness and configuration and general correlations are only suitable for approximate prediction. With this in mind the simple relationship due to Rohsenow and Griffith (1956) is very suitable for practical purposes, and may be expressed in the following form (for a normal gravitational field and adapted to SI units):

$$\frac{q_{crit}}{\rho_g h_{fg}} = 0.0121 \left(\frac{\rho_f - \rho_g}{\rho_g} \right)^{0.6} \text{ m/s} \tag{9.7}$$

For water at atmospheric pressure this becomes:

$$\frac{q_{crit}}{0.597 \times 2256.7} = 0.0121 \left(\frac{957.8 - 0.6}{0.597} \right)^{0.6}$$

$$q_{crit} = 1360 \text{ kW/m}^2$$

which may be compared to the pool boiling critical heat flux indicated in Fig. 9.3.

These correlations may strictly only be applied to pool boiling situations under normal conditions. Boiling which involves specially clean or badly fouled surfaces, finned or specially coated surfaces, liquid metals or high pressures may exhibit rather different characteristics to those described here and the reader is referred to the aforementioned texts for details. A rough indication of the total heat flux in forced convective boiling situations may be obtained by summing the pool boiling heat flux and the forced convection heat flux under similar conditions.

9.2 Condensation

In the discussion of condensation in Section 3.4 two modes of condensation were identified: filmwise and dropwise. It was pointed out that dropwise condensation led to higher heat transfer coefficients than filmwise but could not be maintained in practical applications as the non-wetting agent required for its promotion was eventually washed away. In this section therefore we shall concentrate on filmwise condensation and shall also discuss the detrimental effect of air on the heat transfer coefficient. The following theoretical analysis is essentially due to Nusselt (1916) and leads to an expression which, without the necessity of an empirical constant, predicts fairly accurately the heat transfer coefficient for condensation with laminar condensate flow.

9.2.1 Film condensation on a vertical plate

Consider a vertical plate of unit width arranged so that film condensation occurs on the surface as indicated in Fig. 9.4. The condensing vapour forms a layer of liquid which flows down the plate and increases in thickness as further condensation occurs at the liquid–vapour boundary. This layer of condensate is responsible for the thermal resistance between the condensing vapour and the wall and, as its thickness increases, the resistance increases and the local heat transfer coefficient decreases. In order to determine this thickness, expressions must be derived for three parameters, these being the velocity distribution in the layer, the mass flow rate and the heat flux. An equation for the velocity u as a function of distance y from the surface may be found by resolving forces due to the shear stress τ, gravity and vapour pressure p_g acting on an element of the condensate as indicated in Fig. 9.4. Neglecting momentum effects and assuming the shear stress of the vapour on the flowing liquid is negligible this gives

$$\tau \, dz + \frac{\partial p_g}{\partial z}(\delta - y) \, dz = \rho_f g(\delta - y) \, dz \qquad (9.8)$$

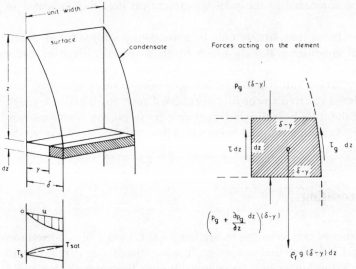

Figure 9.4   Film condensation on a vertical plate.

Expressing τ in terms of the viscosity μ and the velocity gradient:

$$\tau = \mu \frac{du}{dy}$$

and the vapour pressure change across the element ∂p_g in terms of the change in head ∂z:

$$\partial p_g = \rho_g g \, \partial z$$

yields on substitution and rearrangement:

$$\mu \frac{du}{dy} = \rho_f g (\delta - y) - \rho_g g (\delta - y)$$

Integration and substitution of the boundary condition $u = 0$ at $y = 0$ gives

$$u = \frac{(\rho_f - \rho_g)g}{\mu} \left(\delta y - \frac{y^2}{2} \right) \tag{9.9}$$

The incremental mass flow rate of the condensate down the surface is, by continuity

$$d\dot{m} = \rho_f u \, dy \times \text{width}$$

or

$$d\dot{m}' = \rho_f u \, dy \tag{9.10}$$

where the prime on $\dot{m}'$ indicates mass flow per unit width. The total mass flow rate per unit width over the layer of thickness δ is then

$$\dot{m}' = \rho_f \int_0^\delta u \, dy$$

Substitution of equation (9.9) for u and integration yields

$$\dot{m}' = \frac{\rho_f(\rho_f - \rho_g)g}{\mu}\left(\frac{\delta^3}{3}\right) \tag{9.11}$$

The heat flux q to the wall may be expressed as the product of the specific enthalpy difference Δh_s between the vapour and the condensate, and the change of mass of condensate per unit area:

$$q = \Delta h_s \frac{d\dot{m}'}{dz} \tag{9.12}$$

The term Δh_s is composed of two parts: a major part due to the latent enthalpy vaporization h_{fg} and a minor part due to the enthalpy difference between liquid at the saturation temperature T_{sat} (at the free surface where condensation occurs) and the cooler condensate. The specific enthalpy difference may therefore be expressed as

$$\Delta h_s = h_{fg} + \frac{\displaystyle\int_0^{\dot{m}'_\delta} c(T_{sat} - T)\,d\dot{m}'}{\dot{m}'} \tag{9.13}$$

where T is the temperature a distance y from the surface and the integration is carried out over the boundary layer. If the temperature distribution is assumed to be linear it follows, by referring to Fig. 9.4, that

$$\frac{T_{sat} - T}{T_{sat} - T_s} = \frac{\delta - y}{\delta}$$

and

$$T_{sat} - T = \Delta T_s\left(1 - \frac{y}{\delta}\right)$$

where ΔT_s is the conventional condensing temperature difference between the surface and the saturated vapour. Substitution of this expression together with equation (9.10) into equation (9.13) yields

$$\Delta h_s = h_{fg} + \frac{1}{\dot{m}'}\int_0^\delta \rho_f uc\,\Delta T_s\left(1 - \frac{y}{\delta}\right)dy$$

Substitution for u from equation (9.9), evaluation of the integral, substitution of (9.11) for $\dot{m}'$ and algebraic simplification finally gives

$$\Delta h_s = h_{fg} + \tfrac{3}{8}c\,\Delta T_s \tag{9.14}$$

The heat flux to the wall may also be expressed in terms of the conduction through the liquid layer where the temperature distribution is assumed to be linear:

$$q = -k \frac{(T_s - T_{sat})}{\delta} = \frac{k \Delta T_s}{\delta} \tag{9.15}$$

Elimination of q between equation (9.12) and (9.15) gives

$$\frac{d\dot{m}'}{dz} = \frac{k \Delta T_s}{\delta \Delta h_s}$$

In addition $d\dot{m}'$ may be expressed in terms of δ from equation (9.11):

$$d\dot{m}' = \frac{\rho_f(\rho_f - \rho_g)g}{\mu} \delta^2 \, d\delta$$

Elimination of $d\dot{m}'$ and application of the boundary conditions gives

$$\int_0^\delta \delta^3 \, d\delta = \int_0^z \frac{\mu k \Delta T_s}{\rho_f(\rho_f - \rho_g)g \Delta h_s} \, dz$$

Integration then yields an expression for the condensate layer thickness:

$$\delta = \left[\frac{4\mu k \Delta T_s z}{\rho_f(\rho_f - \rho_g)g \Delta h_s} \right]^{1/4} \tag{9.16}$$

Since the local condensation heat transfer coefficient h_z is defined by

$$q = h_z \Delta T_s$$

it follows from equation (9.15) that $h_z = k/\delta$ and thus from equation (9.16)

$$h_z = \left[\frac{\rho_f(\rho_f - \rho_g)g \Delta h_s k^3}{4\mu \Delta T_s z} \right]^{1/4} \tag{9.17}$$

The normal condensation heat transfer coefficient h generally used for practical purposes is an average value over the plate height L (similar to the average h used in forced convection):

$$h = \frac{\int_0^L h_z \, dz}{L}$$

Substitution for h_z gives

$$h = \tfrac{4}{3} h_z \tag{9.18}$$

and

$$h = 0.943 \left[\frac{\rho_f(\rho_f - \rho_g)g \Delta h_s k^3}{\mu \Delta T_s L} \right]^{1/4} \tag{9.19}$$

where

$$\Delta h_s = h_{fg} + \tfrac{3}{8}c \Delta T_s$$

190

This expression agrees well with experimental findings except for the fraction $\frac{3}{8}$ in the term for Δh_s. From experimental data Rohsenow (1956) suggests the amendment

$$\Delta h_s = h_{fg} + 0.68c \, \Delta T_s$$

It is possible for the condensate flow to become turbulent when the surface is long or the condensation rates are high and, as in forced convection, the parameter indicating the commencement of turbulent flow is the Reynolds number conventionally defined as

$$\text{Re} = \frac{\rho U d}{\mu}$$

In this case we are concerned with the flow of condensate down a vertical plate at a mean velocity U and with a mean hydraulic diameter defined (in equation (8.24)) as $4A_c/P$ where A_c is the flow area and P is the 'wetted perimeter'. Thus from continuity

$$\text{Re} = \frac{4\rho U A_c}{\mu P} = \frac{4\dot{m}}{\mu P}$$

where $\dot{m}$ is the condensate mass flow rate. Furthermore, since the wetted perimeter is unity for unit width:

$$\text{Re} = \frac{4\dot{m}'}{\mu} \tag{9.20}$$

for a vertical surface. With the Reynolds number defined in this way the critical value for the transition to turbulent flow is found to be about 1800.

9.2.2 Condensation on tubes

In a conventional surface condenser, condensation occurs on horizontal tubes as indicated in Fig. 9.5a. If the liquid film is thin compared to the tube diameter d the flat plate analysis of the previous section may be applied. The procedure involves replacing g by $g \sin \theta$, where θ is the angle from the top of the tube, and determining a mean value for h by integrating around the tube from $\theta = 0$ to $\theta = 180°$. This yields the expression

$$h = 0.725 \left[\frac{\rho_f (\rho_f - \rho_g) g \, \Delta h_s k^3}{\mu \, \Delta T_s d} \right]^{1/4} \tag{9.21}$$

It is interesting to compare the heat transfer characteristics of a tube in the horizontal and vertical positions. If the tube in the vertical position is considered as a plate of length L, comparison of equation (9.19) and (9.21) yields

$$\frac{h_{\text{horiz}}}{h_{\text{vert}}} = \frac{0.725}{0.943} \left(\frac{L}{d} \right)^{1/4} \tag{9.22}$$

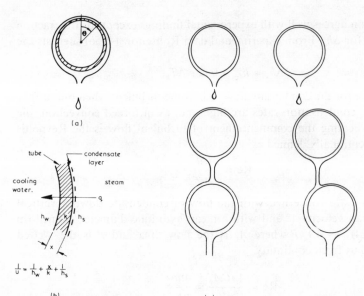

Figure 9.5   Condensing on tubes.

A typical condenser tube will be at least 50 times as long as its diameter and substitution shows that the heat transfer coefficient and therefore the liquid condensed on a horizontal tube will be at least twice that condensed on a similar vertical tube. This assumes of course that other parameters are similar and that the heat transfer is not enhanced by turbulent condensate flow in the case of the vertical tube.

If the tubes in a condenser are arranged in columns as shown in Fig. 9.5c the preceding analysis may only be applied to the top tube as the others have a contribution to the condensate thickness from the tubes above them. Nusselt also analysed the case of a column of horizontal tubes and found that the average heat transfer coefficient for a column of n tubes is given by equation (9.21) modified to

$$h = 0.725\left[\frac{\rho_f(\rho_f - \rho_g)g\,\Delta h_s k^3}{n\mu\,\Delta T_s d}\right]^{1/4} \tag{9.23}$$

In practice equations (9.21) and (9.23) are found to yield rather conservative values of h for water, partly owing to ripples on the condensate film surface, while for liquid metals values tend to be less than those predicted. There is an advantage in staggering condenser tube columns as shown in Fig. 9.5d in order to decrease the proportion of tube covered with condensate from the tube above.

No consideration of condensation, however concise, would be complete without some mention of the effect of non-condensible gas such as air on the heat transfer. In a large power cycle steam condenser operating at low pressure, it is virtually impossible to prevent the ingress of air through leaks. In addition, air dissolved in the boiler water will separate out and eventually

192

pass through to the condenser. The condenser therefore contains a mixture of steam and air. The steam is effectively removed at the condenser tubes so that in the immediate vicinity of the tubes where it is drawn towards them, there is less steam per unit volume than in the bulk of the condenser. The proportion of air to steam is therefore greater near the tube. The effect of this air around the tube is to hinder the mass flow of steam to the condensate layer and thus also hinder the heat flow (as $q = \dot{m}h_{fg}$). A quantity of air in a condenser equal to about 0.5% of the mass of steam may be sufficient to halve the steam-side heat transfer coefficient under some conditions. For this reason an important part of a steam condenser is the air extractor which draws air from the condensing chamber, passes it through a cooler to reduce the quantity of steam extracted and ejects it into the atmosphere (see Fig. 3.5).

Finally the reader is reminded that the steam-side heat transfer coefficient is only one of the parameters involved in heat transfer between the steam and the cooling water, as indicated in Fig. 9.5b. In many condensers the largest thermal resistance is on the cooling water side, and the cooling water flow rate and temperature may be the most important factors determining the condenser performance. For general study of large steam condensers the reader is referred to Anon (1971) and the thermal analysis of condensers given by Silver (1963).

9.3 Free Convection

Free or natural convection is caused by the influence of a body force on a fluid. This body force may be centrifugal or gravitational, or electrical and magnetic in the case of electrically conducting fluids. This brief discussion will be restricted to the most common form of free convection encountered in engineering practice; that is convection due to the density change in a fluid caused by heating or cooling by a submerged surface. The dimensionless groups which are of importance in free convection were introduced in Chapter 3 and are repeated here for convenience. The Grashof number Gr may be considered as representing the ratio of the buoyancy and viscous forces and effectively characterizes free convection in much the same way as the Reynolds number characterizes forced convection. It is defined as

$$Gr = \frac{g\beta \, \Delta T d^3}{\nu^2}$$

where g is the acceleration due to gravity (or other body force) β is the coefficient of volumetric expansion (see Table 9.2 for typical values), ΔT is the temperature difference between the surface and the bulk fluid, d is a linear dimension and ν is the kinematic viscosity. A recent tendency has been to use the Rayleigh number Ra in place of the Grashof number where

TABLE 9.2   The expansion coefficient

Fluid	$T\,(^{\circ}C)$	$\beta \times 10^3\,(K^{-1})$
Air	0	3.92
	20	3.40
	50	3.09
	100	2.68
	150	2.38
	200	2.11
	300	1.75
	400	1.49
	500	1.29
Water	20	0.18
Freon-12	20	2.6
Engine oil	20	~ 0.7
Mercury	20	0.18

$$Ra = \frac{g\beta\,\Delta T d^3 c_p \rho}{vk}$$

$$= Gr\,Pr$$

Many theoretical and empirical free convection relationships can be expressed as:

$$Nu = f(Ra, Pr)$$

or simply

$$Nu = f(Ra)$$

and the Rayleigh number has been found to be a more suitable parameter for indicating the onset of turbulence.

The most frequently studied case of free convection is that of a fluid adjacent to a plane vertical wall. Figure 9.6 shows typical velocity and tem-

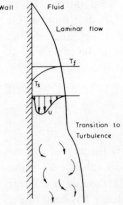

Figure 9.6   Free convection at a cold vertical surface.

perature distributions when the surface is cooler than the surrounding fluid. A description of the theoretical analysis commencing with basic boundary layer equations of the type derived in Section 8.4 is given in Eckert and Drake (1972). An approximate solution is given in Eckert and Gross (1963):

$$Nu = \frac{0.678\ Pr^{1/2}}{(0.952 + Pr)^{1/4}}\ Gr^{1/4}$$

or in terms of the Rayleigh number:

$$Nu = 0.678\left(\frac{Pr\ Ra}{0.952 + Pr}\right)^{1/4} \tag{9.24}$$

where Nu is the mean Nusselt number over the wall height. Exact solutions for the laminar velocity and temperature profiles are given by Ostrach (1952). For the case of vertical surfaces where the linear dimension is the plate height, the flow in the boundary layer becomes turbulent at a Rayleigh number of about 10^9. An analysis of the turbulent boundary layer by Eckert and Jackson (1951) yields

$$Nu = 0.0246\left(\frac{Pr^{1/6}\ Ra}{1 + 0.494\ Pr^{2/3}}\right)^{2/5} \tag{9.25}$$

In both of these equations the dependence of heat transfer on the Prandtl number for gases and many common fluids is small and for practical purposes experimental data are usually expressed in the form

$$Nu = C\ Ra^{m} \tag{9.26}$$

where the Nusselt number is the mean value over the surface and fluid properties are evaluated at the linear mean temperature of the surface and the bulk fluid. Some empirical correlations of this form are included in Table 3.2 and the reader is referred to texts such as McAdams (1954) and Jakob (1949, 1957) for further data.

In recent years some research has been conducted on the important problem of free convection in confined spaces. In the case of horizontal layers free convection only occurs when the upper surface is cooled or when the lower surface is heated. This situation leads to an unstable arrangement in which, at a Rayleigh number (based on distance and temperature difference between the surfaces) in excess of about 1700, a cellular flow pattern occurs. The cells are in the form of hexagonal prisms with their axes vertical and flow is upwards in the interior and downwards at the sides. This manner of flow continues up to a Rayleigh number of about 47,000 above which it changes to an irregular turbulence (Goldstein and Chu, 1969). In the case of vertical layers Elder (1965) has reported that boundary layers develop at each surface with fluid movement upwards at the heated surface and downwards at the cooled surface. In the central portion the temperature distribution is almost linear indicating that heat flow is by conduction. At Rayleigh numbers in excess of about 10^5 horizontal vortices can occur and at values of about 10^9 the flow becomes turbulent.

The calculation of heat flow through a vertical layer of fluid may be based solely on conduction through the fluid at Rayleigh numbers of less than 1000. At higher values a reasonable approximation may be made by assuming each surface exists alone and is surrounded by an infinite amount of fluid at the average temperature of the surfaces. The heat transfer predicted by summing the thermal resistance of each boundary layer in this way has been shown by Kraussold (1936) to be within about 20% of the actual value. Experimental results are sometimes expressed in the form of an equivalent conductivity k_e through a layer of thickness x:

$$q = \frac{k_e}{x} \Delta T_{\text{surfaces}}$$

where q is the total heat transfer and k_e includes the effects of convection, conduction and radiation.

Example 9.1. A domestic hot water radiator is situated in a room at a temperature of 20 °C. The radiator is basically constructed of two pressed steel plates fitted together to form a number of water channels between them and has overall dimensions and heat transfer areas as shown in Fig. 9.7. The

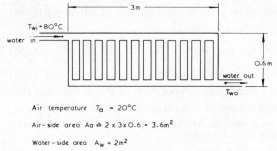

Figure 9.7 The hot water radiator.

air-side heat transfer coefficient h_r due to radiation is 0.005 kW/m² K, the water-side coefficient is 1 kW/m² K and the water mass flow rate and inlet temperature are 0.05 kg/s and 80 °C respectively. Estimate the total heat transfer rate to the room.

Solution. The two sides of the radiator may be treated as two vertical plates for natural convection purposes. Calculation of the Rayleigh number involves a knowledge of the temperature difference between the radiator and the surrounding air. The radiator temperature varies over its length between the inlet and outlet temperature of the water (assuming the radiator metal temperature and the water temperature are virtually the same owing to the high heat transfer coefficient on the water-side compared to the air-side).

Since the water outlet temperature is not given, an estimate must be made and corrected later if found to be considerably in error. For this purpose it is assumed that the water temperature difference across the radiator $(T_{wi} - T_{wo})$ is $8\,°C$ and the mean temperature of the radiator is therefore about $76\,°C$. The required surface to air temperature difference ΔT_m is therefore $76-20 = 56\,°C$ and it should be noted that this value is not very sensitive to error in the estimation of $(T_{wi} - T_{wo})$. The Rayleigh number may now be determined using air properties at the mean of the surface and air temperature, $48\,°C$:

$$\mathrm{Ra} = \frac{g\beta\,\Delta T d^3 c_p \rho^2}{\mu k}$$

$$\mathrm{Ra} = \frac{9.81 \times 3.10 \times 10^{-3} \times 56 \times (0.6)^3 \times 1.01 \times (1.09)^2}{1.96 \times 10^{-5} \times 2.82 \times 10^{-5}}$$

$$= 0.80 \times 10^9$$

The flow is therefore just within the laminar range and the Nusselt number may be determined from equation (9.24) or from the relationship in Table 3.2. Equation (9.24) yields

$$\mathrm{Nu} = 0.678\left(\frac{0.701 \times 0.80 \times 10^9}{0.952 + 0.701}\right)^{0.25}$$

$$= 92$$

and Table 3.2 yields

$$\mathrm{Nu} = 0.59\,(\mathrm{Ra})^{0.25}$$

$$= 102$$

Selection of the latter value obtained from an empirical relationship as the most reliable yields

$$h_a = \frac{k}{d}\,\mathrm{Nu} = \frac{2.82 \times 10^{-5}}{0.6} \times 102$$

$$= 0.0048 \; \mathrm{kW/m^2\ K}$$

(Alternatively the expression in Table 3.3 gives

$$h_a = 0.0014\left(\frac{56}{0.6}\right)^{0.25}$$

$$= 0.0044 \; \mathrm{kW/m^2\ K})$$

Using the estimated value of h_a the total heat transfer may be determined from:

$$Q = UA\,\Delta T_m$$

where

$$\frac{1}{UA} = \frac{1}{h'_a A_a} + \frac{1}{h_w A_w}$$

(In this case it should be noted that the area is included in the summation as it is not the same for the air- and water-sides.) The term h'_a is the total air-side heat transfer coefficient given by the natural convection and radiation components:

$$h'_a = h_a + h_r$$
$$= 0.0048 + 0.005$$
$$= 0.0098 \text{ kW/m}^2 \text{ K}$$

Thus

$$\frac{1}{UA} = \frac{1}{0.0098 \times 3.6} + \frac{1}{1 \times 2}$$
$$UA = 0.0347 \text{ kW/K}$$

and again using the approximate ΔT_m of 56 °C

$$Q = 0.0347 \times 56$$
$$= 1.94 \text{ kW}$$

This heat transfer rate leads to a value of the water temperature difference $(T_{wi} - T_{wo})$ given by

$$Q = \dot{m}c(T_{wi} - T_{wo})$$

where $\dot{m}$ is the water mass flow rate of 0.05 kg/s. Therefore

$$(T_{wi} - T_{wo}) = \frac{Q}{\dot{m}c} = \frac{1.94}{0.05 \times 4.19} = 9.26 \text{ °C}$$

The initial arbitrary estimation of this parameter was therefore sufficiently accurate and, bearing in mind the necessary approximations incurred, the heat transfer rate to the surroundings may be given as about 2 kW.

9.4   The Heat Pipe

The heat pipe is a recently developed device which is capable of transferring large quantities of heat with little associated temperature drop. It consists of a closed pipe containing a working fluid which is boiled at one end and condensed at the other end so that heat is transferred due to the latent heat of vaporization. The liquid which collects at the condensing end is returned to the boiling end by the capillary action in a wick structure fitted to the inside wall of the pipe as shown in Fig. 9.8. This wick is the main feature of the heat pipe and differentiates it from a thermosyphon which utilizes gravity to return the condensate and therefore requires the cool end of the pipe to be above the hot end.

The heat pipe was first patented in America in 1942 but was not developed until the 1960s. It was initially used for devices in spacecraft where it is particularly suitable for operation in the absence of gravity. However, its

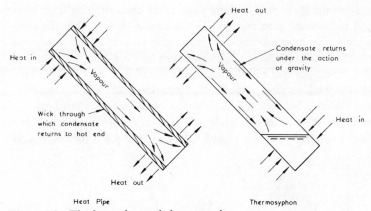

Figure 9.8   The heat pipe and thermosyphon.

application in terrestrial fields, such as the cooling of electronic components, soon became important with the realization of the high heat fluxes that could be obtained. The effective thermal conductivity of heat pipes can be 500 times that of copper of a similar mass, and operating temperatures up to over 1500 °C are possible. Today, many types of heat pipes are available commercially and their applications are widespread (Finlay, 1975). Reviews include Winter and Barsch (1971) and Chisholm (1971) to whom the following simplified analysis is due.

The pressure, and mass flow rate distributions along the length of a heat pipe are shown in Fig. 9.9. The capillary pressure between the ends of the

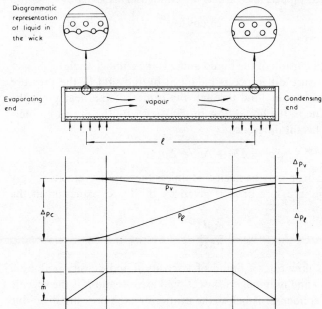

Figure 9.9   Pressure and mass flow rate distribution.

heat pipe leads to the flow of liquid and vapour. In the simplified case considered here it is assumed that there is slightly excess fluid so that the condenser is flooded. The interface between liquid and vapour at the condenser end is outside the wick and the capillary pressure between the phases at this end is zero. In the evaporator end the liquid vaporization causes the interface to recede into the wick leading to an average interface radius of curvature r. The pressure difference between the liquid and vapour in a wick pore may be expressed in terms of an effective wick pore radius r_p. By considering the pore as a short cylindrical tube of radius r_p and applying a force balance as

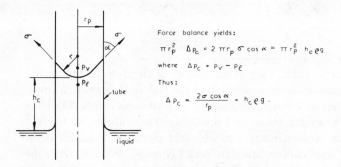

Force balance yields:

$$\pi r_p^2 \ \Delta p_c = 2 \pi r_p \ \sigma \cos \alpha = \pi r_p^2 \ h_c \varrho g$$

where $\Delta p_c = p_v - p_\ell$

Thus:

$$\Delta p_c = \frac{2 \sigma \cos \alpha}{r_p} = h_c \varrho g \ .$$

Figure 9.10 Capillary rise in a tube.

shown in Fig. 9.10 the capillary pressure difference Δp_c at the evaporator (and also for the entire heat pipe as it is zero at the condenser) is given by

$$\Delta p_c = \frac{2\sigma \cos \alpha}{r_p} \tag{9.27}$$

where σ is the surface tension of the liquid and α is the contact angle.

This capillary pressure difference is balanced by the sum of the pressure drop Δp_v due to the vapour flow through the centre of the heat pipe, the pressure drop Δp_1 due to the liquid flow in the wick, and the pressure difference Δp_h due to the height difference:

$$\Delta p_c = \Delta p_v + \Delta p_1 + \Delta p_h$$

The vapour pressure drop may be neglected unless the vapour velocity is very high. The liquid flow pressure drop is given by the Darcy equation in the form

$$\Delta p_1 = \frac{\dot{m}\mu l}{\rho K A}$$

where $\dot{m}$ is the mass flow rate of liquid of viscosity μ and density ρ along a wick of mean length l and nominal cross-sectional area A, and K is the specific permeability. If the condenser is below the evaporator the pressure drop due to the head difference is given by $\rho g l \sin \theta$ where g is the gravitational acceleration and θ is the angle of the heat pipe to the horizontal. Thus the total pressure difference becomes;

$$\Delta p_c = \frac{\dot{m}\mu l}{\rho K A} + g\rho l \sin \theta \tag{9.28}$$

Eliminating Δp_c between (9.27) and (9.28) and rearrangement then yields the mass flow rate:

$$\dot{m} = \frac{AK\sigma\rho \cos \alpha}{\mu l}\left(\frac{2}{r_p} - \frac{\rho g l \sin \theta}{\sigma \cos \alpha}\right) \tag{9.29}$$

The heat flow rate may be found from $Q = \dot{m}h_{fg}$ where h_{fg} is the latent enthalpy of vaporization of the working fluid. Under conditions of horizontal heat pipe operation as shown in Fig. 9.9 ($\sin \theta = 0$) and the common assumption of perfect wetting of the wick ($\cos \alpha = 1$), the heat flow rate becomes

$$Q = \left(\frac{2KA}{lr_p}\right)\left(\frac{\sigma\rho h_{fg}}{\mu}\right) \tag{9.30}$$

The first term on the right-hand side of this equation is a grouping of wick properties only and indicates that a wick structure of high permeability but low pore size is required. These characteristics are to some extent mutually opposed but much progress has been made with wicks of the type shown in Fig. 9.11. The upper pores are small and as it is these pores which mainly determine the capillary effect (being the pores in the neighbourhood of the

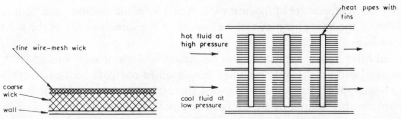

Figure 9.11 Wick arrangement.   *Figure 9.12* Heat exchanger using heat pipes.

interface) this leads to a large value of Δp_c (equation (9.27)). The lower section of wick adjacent to the wall is of coarser structure with greater permeability and leads to a low value of liquid flow resistance. Wicks in commercial heat pipes are commonly constructed from layers of wire mesh or felt held against the side of the pipe by an internal cage or spring. Much research has been conducted on wick arrangements and wick materials such as sintered metals. There is a limiting head over which the heat pipe will operate due to the finite capillary rise height h_c. Under equilibrium conditions this height is given from Fig. 9.10 as

$$h_c = \frac{\Delta p_c}{\rho g}$$

Since h_c can be found experimentally the capillary rise height provides a convenient way of determining Δp_c.

The second term on the right-hand side of equation (9.30) consists entirely of liquid properties and is termed the liquid transport factor N:

$$N = \frac{\sigma \rho h_{fg}}{\mu}$$

It gives an indication of the suitability of possible working fluids for heat pipe application. In addition to a high value of N the fluid must be chemically stable and have a reasonable pressure at the desired operational temperature. For temperatures in the range from room temperature up to about 300 °C, water is one of the most suitable fluids and is often used in commercial heat pipes. A typical tubular heat pipe of 25 mm diameter and 300 mm long, containing water as the working fluid, can transfer a maximum axial heat load of about 5 kW when operating at room temperature.

Operation of a heat pipe at high axial heat load is sometimes hampered by 'dry-out' in the wick of the evaporator section. As the heat to the evaporator end is increased, rapid evaporation and possibly some nucleate boiling occurs and a point is reached where the vapour formed prevents further liquid in the wick from flowing into the evaporator section. The wick is then filled with vapour and the situation is in some ways analogous to 'burnout' in surface boiling. The break-down of the heat pipe can lead to dangerously high temperatures in the evaporator and is generally avoided by operating well below the dry-out temperature of the heat pipe. Apart from the capillary and boiling restrictions the heat pipe may also be limited by the sonic velocity of the axial vapour flow.

In addition to its applications in the space and electronic cooling fields mentioned earlier, the heat pipe has been used in compact heat exchangers and engines. Heat exchangers designed for certain arduous conditions can be considerably simplified by using heat pipes as shown in Fig. 9.12. The exchanger is separated into hot and cool regions and the boundary between the fluids consists of a simple wall, although the surface heat transfer area available to each fluid is large. Only this single wall is required to withstand the stresses caused by the pressure difference between the fluids, and the resulting structural changes may lead to a considerable saving of overall mass or size when compared to a conventional multi-tube and baffle-plate exchanger. An interesting recent application is the 'Vapipe' developed jointly by the National Engineering Laboratory, Glasgow and Shell Petroleum (Lindsay *et al.*, 1970). It consists essentially of a heat pipe situated between the exhaust ducting and carburettor ducting of a normal car engine. Heat is carried by this heat pipe from the exhaust gases to the air and petrol vapour droplet mixture leaving the carburettor, and causes the complete vaporization of the petrol so that a homogeneous petrol–air mixture enters each cylinder. The improved combustion leads to a significant reduction of both fuel consumption and undesirable exhaust emissions. A domestic application is the 'magic cooking pin' consisting of a heat pipe which is inserted into a joint of meat before placing in the oven to roast. Cooking therefore takes place from inside the meat as well as from the outside and the arrangement saves cooking time and reduces energy consumption.

9.5 The Fluidized Bed

The fluidized bed may be visualized as solid particles situated in a vessel which has a base of wire mesh to allow the passage of gas (or liquid) through the particles. At low flow rates the gas flows up through the stationary particles of the bed but at higher flow rates a point is reached where the particles are suspended by the gas. In this fluidized condition the bed of particles is expanded from its original size and the particles move about in a turbulent manner. This turbulence of the particles results in the bed being practically isothermal and leads to high heat transfer coefficients between the bed and immersed surfaces.

The first commercial use of fluidization followed the award of a patent in Germany in 1922 to Fritz Winkler for the gasification of powdered coal. However, full development of the process did not occur until the Second World War when the demand for high octane fuel encouraged the construction of giant fluidized reactors for the catalytic cracking of petroleum. After the war the fluidization process was developed for roasting metal ores, drying powdery materials and for the calcination of limestone. Recent studies of the application of a fluidized bed to the storage of heat from hot gases and to the reduction of iron ore show that fluidization is likely to become of increasing importance. In the remainder of this section we shall limit our discussion to a brief description of fluidization processes and the heat transfer between a fluidized bed and an immersed surface. For further study of heat transfer and other aspects of fluidization the reader is referred to specialized texts such as Kunii and Levenspiel (1969), Davidson and Harrison (1971) and Botterill (1974).

Some of the forms of fluidization found under various conditions are indicated in Fig. 9.13 (a–e). When the flow of fluid (gas or liquid) is low the pressure difference Δp $(p_1 - p_2)$ is insufficient to support the weight of the bed and the fluid percolates through the voids between the particles. When the bed is just supported by the pressure difference it is said to be incipiently fluidized and under this condition a simple force balance yields:

Area × Pressure difference = Weight of Particles and Fluid

$$A_c \, \Delta p = A_c L_i \rho_s (1 - \varepsilon) g + A_c L_i \rho_f \varepsilon g$$

where A_c is the cross-sectional area, L_i is the height of the bed at incipient fluidization, ε is the voidage volume, ρ_s and ρ_f are the densities of the solid

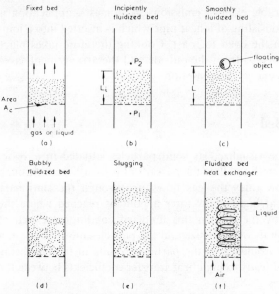

Figure 9.13 Forms of fluidization.

and fluid and g is the gravitational acceleration. In gas fluidized beds the fluid term may be neglected to give

$$\Delta p = L_i \rho_s (1 - \varepsilon) g$$

Further increases in fluid flow may lead either to a smoothly fluidized bed (normally found when the fluid is a liquid) or to a bubbling fluidized bed (normally found with a gas). In both of these conditions the bed will support light objects or flow from vessel to vessel to maintain equal height and generally exhibit the characteristics of a liquid. A less desirable form of fluidization is termed 'slugging' and involves an unstable situation in which larger amounts of solid particles are caught above the rising gas. Finally, at still higher flow rates, particles will be carried out of the bed by entrainment in the fluid.

The heat transfer between the fluid passing through a fluidized bed and the bed is usually total; that is to say the fluid and the particles reach the same temperature within the bed. Furthermore, the circulation and turbulence in a fully fluidized state is sufficient to maintain the entire bed at a uniform temperature. The main resistance to heat transfer in a fluidized bed, therefore, occurs between a solid surface and the bed. In the case of heat transfer between air and liquid as shown in Fig. 9.13f for example, there is a heat transfer resistance between the bed and the coiled pipe (and also between the liquid and inner wall of the pipe) but virtually no resistance between the air and the bed. We shall now consider the heat transfer between a surface and a fluidized bed.

The typical variation between surface heat transfer coefficient and fluid velocity through the bed is shown in Fig. 9.14. The pressure difference across the bed is also included because in many situations it is desirable to transfer the maximum heat with the minimum possible pressure drop and therefore the minimum pumping power. The peak heat transfer coefficient between a gas fluidized bed and an immersed surface is about 0.1 to 0.5 kW/m² K which is up to an order of magnitude greater than the coefficient obtained under similar conditions of gas flow in forced convection.

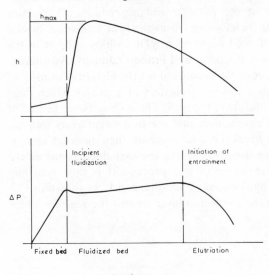

Figure 9.14 Variation of surface heat transfer coefficient and pressure difference with fluid velocity.

There are a number of correlations available in the literature for the heat transfer rate under conditions of full fluidization. Most of the relationships are semi-empirical and based on dimensional analysis of the variables involved. For estimation of heat flow from the container walls to the bed one of the simpler expressions is due to Levenspiel and Walton (1954):

$$\frac{hd_p}{k} = 0.6\left(\frac{d_p\rho u}{\mu}\right)^{0.3}\left(\frac{c_p\mu}{k}\right)$$

where properties refer to the fluid and d_p is the mean particle diameter. Heat flow to horizontal tubes in a fluidized bed was correlated by Vreedenberg (1960) as:

$$\frac{hd_t}{k} = 0.66\left[\left(\frac{d_t\rho u}{\mu}\right)\left(\frac{\rho_s}{\rho}\right)\left(\frac{1-\varepsilon}{\varepsilon}\right)\right]^{0.44}\left(\frac{c_p\mu}{k}\right)^{0.3}$$

where d_t is the tube diameter and ρ_s is the solid particle density. This expression is applicable when the Reynolds number based on the bulk velocity u is less than 2000:

$$\frac{d_t\rho u}{\mu} < 2000$$

For higher flow rates the experimental data is correlated by

$$\frac{hd_t}{k} = 420\left[\left(\frac{d_t\rho u}{\mu}\right)\left(\frac{\rho_s}{\rho}\right)\left(\frac{\mu^2}{d_p^3\rho_s^2 g}\right)\right]^{0.3}\left(\frac{c_p\mu}{k}\right)^{0.3}$$

One way of accounting for the high heat transfer coefficients in a fluidized bed is to consider the scouring action of the particles on an immersed surface. The violent turbulence of the particles and air reduces the effective boundary layer thickness and therefore increases the heat transfer. Correlations are similar in form to forced convection relationships and, as in the cases above, involve the Nusselt, Reynolds and Prandtl numbers. An alternative mechanism which has successfully accounted for the high heat transfer in some situations involves consideration of clusters of particles which continuously make their way to and from the wall. These clusters are heated at the wall, mainly by unsteady conduction, and are then swept away into the bulk of the bed where they break up and dissipate their internal energy. Visualization studies and measurements of the thermal fluctuations at the surface have confirmed the existence of this process. It is now generally accepted that both these mechanisms occur simultaneously as shown in Fig. 9.15. Adjacent to the surface there is a thin layer termed the gas film where

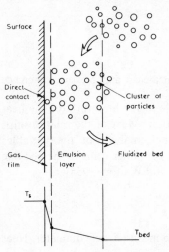

Figure 9.15   Boundary layer in a fluidized bed.

thermal resistance is mainly due to conduction through the gas (enhanced to some extent by occasional direct contact of the particles with the surface). Between this film and the bulk of the bed there is an emulsion layer characterized by the constant interchange of clusters of particles. One of the successful aspects of this analysis is that it accounts for the observation that the heat transfer coefficient is higher for small areas (where the emulsion layer is not properly developed) than for large areas.

Summaries, Symbols, Units and Problems 10

10.1 General Comments

This chapter consists of general items which could have been distributed throughout the rest of the book. It seemed unfortunate, however, to constantly distract the readers attention with items of 'book-keeping' such as summaries, symbols, units and problems. They are therefore collected together at the end and distributed under the general headings of radiation, conduction, convection and heat exchangers. No further reference is made to Chapters 4 and 9 as they are already fairly condensed and self-contained.

The summaries basically consist of identified equations from the text and it is assumed that the text will be consulted if information on the limitations of use of the equations is required. The symbols used are carefully identified where first encountered and at other strategic points in the text. Notation listed in this section consists of general symbols and not those used only in particular derivations or examples. As a general policy, conventional notation has been used in spite of the fact that this has resulted in a dual meaning for a few symbols. The units conform to the Système International d'Unités although occasional infiltration of the bar or centimetre is allowed. The problems are roughly divided under three headings and while some are trivial, the majority are designed to be completed in about half an hour each once the relevant subject matter has been understood. Tables (such as Mayhew and Rogers, 1969) giving properties of water, steam and air are required for the solution of some problems. A few relevant conversion factors and material properties are presented following the end of this chapter.

10.2 Radiation (Chapters 1 and 6)

10.2.1 Summary

Stefan–Boltzmann law

$$E_b = \sigma T^4 \tag{1.1}$$

Emissivity

$$\varepsilon = E/E_b \tag{1.2}$$

For solids

$$\alpha + \rho = 1 \ (\text{if } \tau = 0) \tag{1.4}$$

$$\alpha = \varepsilon$$

For a blackbody

$$\alpha = \varepsilon = 1$$

$$\rho = 0$$

Radiosity B and irradiation H

$$B = \rho H + \varepsilon E_b \tag{1.5}$$

Net heat flow from a surface

$$\frac{Q}{A} = B - H \tag{1.6}$$

Heat flow from surface 1 to black or distant grey surroundings 2

$$Q = -\sigma \varepsilon_1 A_1 (T_2^4 - T_1^4) \tag{1.7}$$

Heat flow between black surfaces

$$Q = -\sigma A_1 F_{12}(T_2^4 - T_1^4) \tag{1.10}$$

Shape factor F_{12} of surface 2 viewed from surface 1

$$A_1 F_{12} = \int\limits_{A_1} \int\limits_{A_2} \frac{\cos \phi_1 \cos \phi_2 \, dA_1 \, dA_2}{\pi r^2} \tag{1.16}$$

Shape factor reciprocity

$$A_1 F_{12} = A_2 F_{21} \tag{1.17}$$

Radiation between two black spheres a large distance x apart

$$Q = -\frac{\pi r_1^2 r_2^2 \, \sigma(T_2^4 - T_1^4)}{x^2} \tag{Section 6.1}$$

Heat transfer coefficient

$$Q = -h_r A_1 (T_2 - T_1)$$

Parallel grey surfaces of infinite extent

$$Q = \frac{-\sigma A(T_2^4 - T_1^4)}{1/\varepsilon_1 + 1/\varepsilon_2 - 1} \tag{6.3}$$

Radiation shields—all emissivities equal

$$\frac{Q_n}{Q_0} = \frac{1}{n+1} \tag{Section 6.1}$$

$$Q_n = \text{radiation with } n \text{ shields}$$

$$Q_0 = \text{radiation with no shields}$$

Enclosed space with n plane sides

$$F_{12} + F_{13} + F_{14} + \ldots F_{1n} = 1$$

Radiation network analogy

$$\text{Shape resistance} = \frac{1}{A_1 F_{12}} \tag{1.18}$$

$$\text{Surface resistance} = \frac{1 - \varepsilon}{\varepsilon A} \tag{1.19}$$

Tables and charts

Emissivity of surfaces at room temperature	Table 1.1
Shape factor charts:	
Rectangles at right angles	Figure 6.6
Parallel rectangles	Figure 6.7
Parallel discs	Figure 6.12b
General network for two radiating surfaces in distant grey surroundings	Figure 6.12c

10.2.2 Symbols and units (Chapters 1 and 6)

A	Surface area (m^2)
B	Radiosity—total radiation leaving a surface (kW/m^2)
c	Velocity of light (m/s)
E	Emissive power (kW/m^2)
E_b	Emissive power of a black surface (kW/m^2)
$E_{b\lambda}$	Monochromatic emissive power of a black surface (kW/m^2–μm)
F_{12}	Shape factor of surface 2 viewed from surface 1
H	Irradiation—total incident radiation (kW/m^2)
h	Planck's constant (kJ–s)
h_r	Radiation heat transfer coefficient (kW/m^2 K)
I_n	Normal intensity of radiation (kW/m^2)
I_ϕ	Intensity of radiation at angle ϕ (kW/m^2)

m	Mass (kg)
n	Number of radiation shields
Q	Heat flow rate (kW)
q	Heat flow rate per unit area (kW/m^2)
r	Radius (m)
T	Temperature (°K)
x	Distance (m)
α	Alpha—Absorptivity
ε	Epsilon—Emissivity
λ	Lambda—Wavelength (μm)
ν	Nu—Frequency (s^{-1})
ρ	Rho—Reflectivity
σ	Sigma—Stefan–Boltzmann constant (kW/m^2 K^4)
τ	Tau—Transmissivity
ϕ	Phi—Angle to normal (degrees)

10.2.3 Problems

(Note $\sigma = 56.7 \times 10^{-12}$ kW/m^2 K^4.)

1. A water tank measures 2 m $\times$ 1 m by 1 m high and radiates from the top and each side. The surface emissivity and temperature are 0.9 and 25 °C and the temperature of the surroundings is 2 °C. Estimate the heat lost by radiation from the tank and state the assumptions made in the estimation. What is the reduction in heat loss if the tank is coated with aluminium paint of emissivity 0.5?
(0.89 kW, 0.39 kW)

2. What are radiation configuration factors or shape factors and why are they used?
   A room measuring 3 m $\times$ 4 m $\times$ 2$\frac{1}{2}$ m high has the ceiling covered with heating panels. Under steady-state conditions the ceiling is at a temperature of 50 °C and the walls and floor at a temperature of 20 °C. Assuming all the surfaces have an absorptivity of unity calculate the net heat radiated from the ceiling using configuration charts.
   Check the result obtained by calculating the net radiation from the ceiling to blackbody surroundings.
(2.37 kW)

3. A spherical satellite of 1 m diameter encircles the earth at an altitude of 300 miles. Estimate the shape factor of the earth from the satellite and hence calculate the equilibrium temperature of the satellite on the 'dark' side and on the 'bright' side of the earth. Assume the earth is 8000 miles diameter and has a blackbody temperature of 20 °C. The temperature of outer space may be taken as 0 °K and the satellite is irradiated with a heat flux of approximately 1.3 kW/m^2 from the sun when on the bright side. (1 mile = 1610 m.)
(147 °K, 297 °K)

4. A room has a radiating wall panel fitted along the entire length of one wall. The panel extends from the ground to a height of 1 m and has a surface temperature and emissivity of 60 °C and 0.9 respectively. The room measures 4 m × 4 m × 2½ m high, and the walls are effectively black with a temperature of 15 °C. Calculate the heat radiated to the floor and the ceiling.
(0.41 kW, 0.17 kW)

5. A cube of side 10 mm has a surface emissivity of 0.9 and is situated 100 mm from an electric arc. For radiation purposes the arc may be considered as a black sphere of 5 mm diameter at a temperature of 3727 °C. The surroundings are at an ambient temperature of 27 °C. Estimate the equilibrium temperature of the cube in each of the following situations:

  (a) as shown in the sketch

  (b) as in the sketch but rotated 45° about the centre as shown

  (c) as in position *a*, but with the arc moved vertically downwards by 50 mm and all sides of the cube, except that side facing the arc, thermally insulated.

(159 °C, 181 °C, 318 °C)

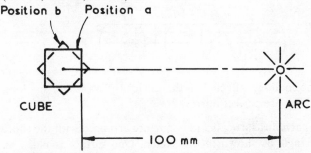

6. A blackbody in the form of a disc is situated in a plane which is 0.5 m from an electric arc as shown in the sketch. The heat energy received from the arc when the mirror is not present causes the temperature of the disc to rise to a steady value of 0.5 °C above the surroundings at 26.5 °C. The total heat transfer coefficient between each side of the disc and the surroundings is

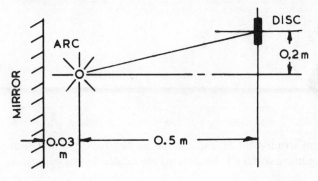

0.1 kW/m² K. If the arc may be considered as a black sphere of 6 mm diameter, estimate the temperature of the arc.

A mirror with perfect reflectivity is placed a distance of 0.03 m behind the arc as shown. Estimate the new steady temperature of the disc.
(1710 °C, 27.4 °C)

7.   If the radiation shape factor between two adjacent sides of a hollow cube is 0.2 determine the shape factor between two opposite sides.

One side of a large hollow cube of side length 2 m is black and has a temperature of 200 °C and the opposite side is black with a temperature of 0 °C. The other sides are grey with an emissivity of 0.5 and a temperature of 30 °C. Determine the radiation heat transfer (from all the surfaces) to the side opposite the black surface.
(0·2, 5kW)

8.   Design a radiation network analogue for two grey surfaces which radiate to each other in a thermally insulated enclosure. Show that when the surface areas are unity the net radiation between the surfaces is given by the following expression, where the symbols have the usual meaning:

$$q = \frac{\sigma A(T_1^4 - T_2^4)}{\left(\dfrac{2}{1 + F_{12}}\right) + \left(\dfrac{1 - \varepsilon_1}{\varepsilon_1}\right) + \left(\dfrac{1 - \varepsilon_2}{\varepsilon_2}\right)}$$

In what way is the analogy affected if the enclosure walls are no longer thermally insulated but are perfect mirrors?

9.   Three surfaces each measuring 0.2 m × 5 m are arranged with the shorter sides forming a triangle as shown in the sketch. One surface is black at a temperature of 200 °C, one is grey with an emissivity of 0.5 and a temperature of 100 °C and the third is a perfect reflector. Neglecting the heat radiated to the surroundings from the ends of the surfaces, determine the heat transfer rate from the black to the grey surface.
(0.74 kW)

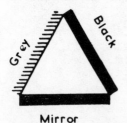

Mirror

10.   Define the terms irradiation H, radiosity B and emissivity ε and show that for a non-transmitting surface 1 these terms are related by the expression

$$H = \frac{B - \varepsilon_1 E_{b1}}{1 - \varepsilon_1}$$

where E_{b1} is the blackbody emissive power of surface 1.

Noting that the heat flux leaving a surface is given by $q = B - H$ show how radiation between surfaces may be estimated by means of a network involving surface and space resistances.

An electric arc, which may be considered for radiation calculations as a black sphere of 5 mm diameter at a temperature of 5000 °C, is situated 250 mm from a metal disc of 100 mm diameter. The disc is oriented at right angles to the normal to the arc and is open to radiation on both sides. Assuming heat exchange with the disc is entirely by radiation, estimate the temperature of the disc when the room temperature is 20 °C.
(147 °C)

11. Two parallel plates measuring 1 m × 1 m are spaced 2 m apart. The inner surface of each plate radiates as a blackbody and the outer surfaces are perfectly insulated. A radiation shield with an emissivity of 0.05 each side and measuring 1 m × 1 m is situated equidistant between the plates. The temperature of one plate is maintained at 727 °C, the other plate at 227 °C and the surroundings are at 27 °C. Sketch the radiation network and calculate the heat transfer from the hot plate. (The configuration factor for two square parallel planes at a distance apart equal to one of the sides is 0.2.)
(54 kW)

12. Derive an expression for the radiation heat transfer between two parallel grey surfaces either by using the network analogy or by summation of successive reflections.

Air flows at the rate of 20 m/s between two parallel plates measuring 1 m × 1 m. One plate has an emissivity of 0.9 and is maintained at a temperature of 300 °C and the other plate has an emissivity of 0.5, a temperature of 200 °C and is perfectly insulated. The heat transfer by convection to the air is given by the relationship

$$Nu = 0.036 \, Re^{0.8} \, Pr^{0.33}$$

where the linear dimension is the plate length. By making reasonable assumptions estimate the mean air temperature and the total heat flow rate to the air.
(172 °C, 8.53 kW)

13. A hollow cubic box of side length 1 m has radiation black internal surfaces. The base has part of the internal surface covered with a thin black copper plate. This plate is fitted in a corner and measures 0.5 m × 0.5 m, thus covering a quadrant of the base. The plate temperature is 327 °C and all the other surfaces are at a temperature of 127 °C. Determine the radiation heat transfer from the plate to each of the five other sides.

14. The hollow box with a plate covering a quadrant of the base as described in problem 13 has the lid removed so that the top is open to distant sur-

roundings at a temperature of 27 °C. Assuming the plate and internal side temperatures remain at 327 °C and 127 °C, estimate the total radiation heat flow rate to the surroundings.

15.   Three square surfaces each of 1 m side length are arranged as shown in the sketch and have the indicated temperature and surface characterstics. Assuming there is no heat flow from the backs of the plates estimate the total heat flow rate by radiation from the plates to the surroundings at a temperature of 27 °C.

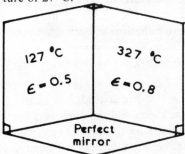

16.   A furnace, similar to that shown in Fig. 6.11, has the recess (surface 1) at a temperature of 1000 °C, the back (surface 1′) at 800 °C, the front (surface 3) at 400 °C and the four sides at 600 °C. The recess is effectively a black area but all the other surfaces have an emissivity of 0.8. Determine the total heat transfer to the front (surface 3). The radiation network may be considerably simplified by lumping the four sides together and treating them as a single surface and therefore a single node in the network.

17.   Solid mercury in the form of a cube of side 10 mm is attached to a cord and is lowered into a vacuum chamber. The chamber walls are radiation black at a temperature of 20 °C and the mercury block is initially 10 degrees below the melting point of mercury (of 234.3 °K). The emissivity, specific heat and density of the mercury may be taken as 0.3, 0.145 kJ/kg K and 13.7×10^3 kg/m^3 respectively.

Derive the differential equation for the time taken for the mercury to reach the melting point and hence by suitable simplification estimate this time. Would the block melt immediately and flow away to the base of the chamber at the end of this period?
(7.1 min)

10.3   Conduction (Chapters 2 and 7)

10.3.1   Summary

Conduction equation in one dimension

$$Q = -kA \frac{(T_2 - T_1)}{x} \tag{2.1}$$

Radial flow through a tube

$$Q = -\frac{2\pi kl(T_2 - T_1)}{\ln(r_2/r_1)} \tag{2.15}$$

Composite wall

$$Q = -UA\,\Delta T_{\text{overall}} \tag{2.4}$$

where

$$\frac{1}{U} = \sum\frac{1}{h} + \sum\frac{x}{k} \tag{2.6}$$

Composite tube

$$\frac{Q}{l} = U'\,\Delta T_{\text{overall}} \tag{2.16}$$

where

$$\frac{1}{U'} = \sum\frac{1}{2\pi hr} + \sum\frac{\ln(r_{\text{out}}/r_{\text{in}})}{2\pi k} \tag{2.17}$$

Thermal conductivity

solids

$$k = k_{\text{ph}} + k_{\text{e}} \tag{2.7}$$

$$k_{\text{e}} = \sigma LT \tag{2.9}$$

liquids

$$k = k_{\text{ph}} + k_{\text{t}} \tag{2.10}$$

Conduction equation in three dimensions—constant k

$$\frac{\partial^2 T}{\partial x^2} + \frac{\partial^2 T}{\partial y^2} + \frac{\partial^2 T}{\partial z^2} + \frac{q_{\text{g}}}{k} = \frac{1}{\alpha}\frac{\partial T}{\partial t} \tag{7.3}$$

Conduction equation in radial coordinates—constant k and axisymmetrical.

$$\frac{\partial^2 T}{\partial r^2} + \frac{1}{r}\frac{\partial T}{\partial r} + \frac{\partial^2 T}{\partial z^2} + \frac{q_{\text{g}}}{k} = \frac{1}{\alpha}\frac{\partial T}{\partial t} \tag{7.6}$$

Unidimensional steady-state conduction:
with no heat generation

$$\frac{\partial^2 T}{\partial x^2} = 0 \tag{7.8}$$

$$T = Mx + N$$

with heat generation

$$\frac{\partial^2 T}{\partial x^2} = -\frac{q_{\text{g}}}{k} \tag{7.9}$$

$$T = -\frac{q_{\text{g}}}{k}\frac{x^2}{2} + Mx + N$$

if q_{g} electrical, $q_{\text{g}} = \left(\frac{I}{A}\right)^2 \rho$ \tag{7.12}

215

with no heat generation but convection from sides

$$\frac{\partial^2 \Delta T}{\partial x^2} = \frac{hP}{kA} \Delta T \tag{7.13}$$

$$\Delta T = M\,e^{-mx} + N\,e^{mx} \tag{7.14}$$

where

$$\Delta T = T - T_s$$

$$m = \sqrt{\frac{hP}{kA}}$$

fin efficiency $\eta_f = \dfrac{\text{actual heat transferred}}{\text{heat transferred if entire fin area is at the}}$
$\phantom{\text{fin efficiency } \eta_f = \dfrac{}{}}$ temperature of the base of the fin

Multidimensional steady-state conduction
Curvilinear squares

$$Q = -\frac{N}{I}\,wk(T_2 - T_1) \tag{7.16}$$

Relaxation of node temperatures

The residual temperature at a node is the summation of the temperatures of the 4 surrounding positions less 4 times the temperature of the node.

Unsteady-state heat flow to a body with a uniform internal temperature
Body in infinite reservoir of fluid

$$\frac{T_b - T_f}{T_{b0} - T_f} = \exp\left[-\left(\frac{hA}{mc}\right)_b t\right] \tag{7.26}$$

Body in small mass of fluid

$$\frac{T_b - T_\infty}{T_{b0} - T_\infty} = \exp\left[-h_b A_b\left(\frac{1}{m_b c_b} + \frac{1}{m_f c_f}\right)t\right] \tag{7.33}$$

Unsteady-state conduction
Step change in temperature at surface of semi-infinite plate

$$\frac{T - T_s}{T_i - T_s} = \mathrm{erf}\left(\frac{x}{2\sqrt{\alpha t}}\right) \tag{7.41}$$

Dimensionless number representation

$$\theta = f(\mathrm{Bi},\ \mathrm{Fo})$$

Schmidt graphical method

$$\Delta t = \frac{(\Delta x)^2}{2\alpha} \tag{7.46}$$

Tables and charts

10.3.2 Symbols and units (Chapters 2 and 7)

A Area (m^2)

a Molecular spacing

Bi   Biot number

c Specific heat at constant pressure (kJ/kg K)

c_v Specific heat at constant volume (kJ/kg K)

d Molecular diameter

Fo   Fourier number

h Heat transfer coefficient (kW/m^2 K)

I Number of temperature increments (curvilinear squares)

I Electrical current (amps)

i Electrical current density (amps/m^2)

K_0 Boltzmann constant (kJ/K)

k Thermal conductivity (kW/m K)

k_e Electronic component of thermal conductivity (kW/m K)

k_{ph} Phonon component of thermal conductivity (kW/m K)

k_t Translational component of thermal conductivity (kW/m K)

L Lorenz number (2.45×10^{-8} WΩ/K^2)

L Length (m)

l Length (m)

M Molecular mass or molecular 'weight'

m Mass (kg)

N Molecules per mole or Avogadro's number

N Total number of heat flow tubes (curvilinear squares)

P Wetted perimeter (m)

Q Heat flow rate (kW)

Q_g Heat generation rate within a material (kW)

q Heat flow rate per unit area (kW/m^2)

q_g Heat generation rate per unit volume (kW/m^3)

R Electrical resistance (Ω)

R_0 Universal gas constant (8.314 kJ/kg-mole K)

r Radius (m)

T Temperature (°C or °K)

T_b Temperature of body (°C)

T_{b0}	Temperature of body at time $t = 0$ (°C)
T_f	Temperature of fluid (°C)
T_i	Initial temperature (°C)
T_s	Temperature of surroundings (°C)
T_∞	Final or steady-state temperature (°C)
ΔT	Temperature difference (°C)
t	Time (s)
t	Thickness (m)
Δt	Time increment (graphical method) (s)
U	Overall heat transfer coefficient (kW/m² K)
U'	Overall heat transfer coefficient based on unit length of tube (kW/m² K/m)
U_i	Internal energy (kJ)
u	Specific internal energy (kJ/kg)
v	Average phonon velocity (m/s)
w	Width (m)
w	Mesh spacing (conduction analogue) (m)
x	Distance (m)
Δx	Distance increment (graphical method) (m)
α	Alpha—Thermal diffusivity, $k/\rho c$ (m²/s)
η	Eta—Dummy variable
η_f	Eta—Fin efficiency
θ	Theta—Dimensionless temperature difference
λ	Lambda—Mean free path of particles
λ	Lambda—Constant
ρ	Rho—Density (m³/kg)
ρ	Rho—Electrical resistivity (Ω-cm)
σ	Sigma—Electrical conductivity (Ω-cm)$^{-1}$
μ	Mu—Dynamic viscosity (kg/ m s)

10.3.3 Problems

1. The door of a domestic refrigerator has an area of 0.7 m² and basically consists of a thin metal sheet with a 25 mm thick layer of insulation on the inside. The thermal conductivity of this insulation is 0.25×10^{-3} kW/m K and the heat transfer coefficients to the surrounding air each side of the door are both 0.01 kW/m K. Determine the heat flow rate through the door and the temperature of the metal sheet under conditions when the cold chamber and room temperatures are 0 °C and 20 °C respectively.
(46.7 W, 13.3 °C)

2. A tube within a heat exchanger has an internal diameter of 25 mm and a wall thickness of 2 mm. Gas at a temperature of 500 °C flows around the outside of the tube and liquid at 400 °C flows through the inside. The outer and inner heat transfer coefficients are 0.2 kW/m² K and 0.5 kW/m² K.

Determine the heat flow rate per metre length to the liquid under conditions when the tube is constructed of:

(a) ceramic material with a thermal conductivity of 1×10^{-3} kW/m K

(b) mild steel with a thermal conductivity of 45×10^{-3} kW/m K

In case (a) determine the error involved in treating the problem as conduction through a plate of length equal to the mean perimeter of the tube.
(9.6 kW/m, 12.4 kW/m, 2%)

3.   A long copper rod of 30 mm diameter conducts a current of 1000 amps and has an electrical resistance of 20×10^{-6} Ω per metre length. The rod is insulated with a layer of fibrous cotton, of thermal conductivity 0.06 W/m K and outer diameter 36 mm, covered by a layer of plastic of thermal conductivity 0.4 W/m K. The surroundings are at a temperature of 20 °C and the heat transfer coefficient between the plastic and the surroundings is 20 W/m² K.

Determine the thickness of the layer of plastic which gives the minimum temperature in the cotton insulation. For this condition find the nominal temperature of the copper rod and the maximum temperature in the plastic layer.

(2 mm, 38.4 °C, 28.8 °C)

4.   Common insulation materials are designed primarily to divide the air contained within them into small pockets. For this reason good insulation materials have a conductivity which approaches that of air. With this in mind explain why the conductivity of microporous silica is less than that of air (when situated in air and other conditions being equal).

Design in some detail an experimental arrangement suitable for measuring the conductivity of a good insulation material.

5.   A steel rod of 50 mm diameter and 1 m long is connected at one end to a large vessel which has a constant temperature of 154 °C. The other end of the rod is insulated and has a temperature of 42 °C. Assuming that the surroundings are at 20 °C and the thermal conductivity of the steel is 45 W/m K estimate the heat transfer coefficient between the rod and the surroundings.
(3.5 W/m² K)

6.   A metal rod projects from a heat source which is maintained at a constant temperature. The sides and free end of the rod are covered with a thin layer of insulation of thickness i and conductivity k_i. There is no heat flow from the end of the rod to the surroundings.

Derive an equation for the temperature distribution along the rod and write down the boundary conditions. How is the solution affected if the insulation at the end of the rod is removed?

7.   A rod of plutonium-238 with a power density of 0.563 W/g and of length 100 mm is surrounded by an annulus of thermoelectric elements and an annulus of thermal insulation as shown in the sketch. The annulus of

thermoelectric elements has an outer radius of 20 mm and outer temperature of 100 °C and a thermal efficiency of 5 %. The insulation has a thermal conductivity of 0.4 W/m K and the heat transfer coefficient between the outer surface of the insulation and the surrounding air (at a temperature of 20 °C) is 10 W/m² K. It may be assumed that the ends of the generator are perfectly thermally insulated.

Determine the outer radius of the insulation which will maximize the temperature difference across the thermoelectric elements. At this optimum insulation thickness estimate the electrical power output and the mass of plutonium required.
(40 mm, 0.62 W, 22 g)

8. The effectiveness of a fin in transferring heat to the surroundings is indicated by a parameter termed the 'fin efficiency' and defined as the ratio of the actual heat transferred to the heat which would be transferred if the entire fin area were at the base temperature of the fin.

Derive an expression for the fin efficiency of a circular rod, length l, which projects from a heat source at temperature T_1 into surroundings at T_s. It may be assumed that heat flow from the side of the rod is determined by a constant heat transfer coefficient h and heat flow from the end of the rod is negligible.

9. A large plate has a uniform thickness l and a thermal conductivity k. The wall temperature on both surfaces of the plate is T_w and heat is generated electrically within the plate at the rate of $\dot{q}_g$ per unit volume. Derive an expression for the maximum temperature within the plate and also for the heat flux at the wall.

10. Current I flows along an electrical conductor, resistivity ρ of square cross-section, side a. The top and bottom surfaces of the conductor are insulated and the I^2R heating is dissipated from the sides which are at a temperature T_s. Obtain the steady heat conduction equation for the con-

ductor and hence show that the maximum temperature T_{max} in the conductor of thermal conductivity k is given by:

$$T_{max} = T_s + \frac{I^2 \rho}{8ka^2}$$

11. A bar of square cross-section connects two metallic structures. One structure is maintained at a temperature of 200 °C and the other is maintained at 50 °C. The bar is 100 mm long, has a cross-section of 20 mm × 20 mm and is constructed of mild steel with a thermal conductivity of 0.06 kW/m K. The surroundings are at a temperature of 20 °C and the heat transfer coefficient between the bar and the surroundings is 0.01 kW/m² K.

Derive an equation for the temperature distribution along the bar and hence calculate the total heat flow rate from the bar to the surroundings.
(7.6 W)

12. Two thermal reservoirs, maintained at a constant and equal temperature are connected by a metal rod. Heat is generated by the passage of current in the rod (which is not insulated). Derive from first principles an expression for the temperature difference between the mid-point of the rod and the reservoirs State any assumptions made.

13. A tube constructed of ceramic material with a thermal conductivity of 1 watt/m K has a square section with an internal side of 40 mm and an external side of 80 mm. The inner wall temperature is 50 °C and the outer wall temperature is 100 °C. By constructing curvilinear squares (or by some other method) estimate the heat flow rate from the inner to the outer surface per unit length of tube.

The same tube is used in a heat exchanger in which, at a certain section, air flowing through the centre is at 100 °C and air flowing around the outside is at 200 °C. If the heat transfer coefficients are both equal to 0.05 kW/m² K calculate the heat exchange rate between the air flows per unit length of tube at this section.
(0.5 kW, 0.17 kW)

14. The chimney of a gas-heated furnace has a cross-section with outer measurements of 1.2 m × 0.9 m. The walls are 0.3 m thick and the inner duct measures 0.6 × 0.3 m. The inner surface has a temperature of 700 °C, the outer surface a temperature of 100 °C and the wall conductivity is 1 W/m K. Estimate the heat flow rate from the chimney to the surroundings per metre length under these conditions by using:

(a) Curvilinear squares.

(b) Numerical analysis.

(5.2 kW)

15. The temperature of air in a duct is measured by means of a thermo-

couple. The sensing end of the couple consists of two wires, 0.5 mm diameter, which project 10 mm into the duct and are welded together at the tips to form a hot junction. Both wires have a thermal conductivity of 0.02 kW/m K. Derive an equation for the temperature distribution along the wires.

If the duct wall temperature is 100 °C, the air temperature, pressure and velocity are 20 °C, 1 bar and 5 m/s respectively, determine the error in the air temperature as indicated by the thermocouple. Convective heat transfer between couple and air stream is described by the relationship $Nu = 0.7 Re^{0.5}$; the Reynolds number is based on the wire diameter. Radiation may be neglected. (The temperature distribution equation may be simplified considerably if it is assumed that the measurement error is small.)
(1.08 °C)

16. A spherical metallic meteorite of 2 cm diameter enters the earth's atmosphere at a temperature of 20 °K. The mean rate of heat transfer to the meteorite is 300 kW/m². The mean specific heat and density are 0.5 kJ/kg K and 8000 kg/m² and the temperature and heat of sublimation are 3000 °C and 250 kJ/kg. Estimate approximately how long the meteorite lasts after entering the atmosphere.
(2.85 min)

17. When is it permissible to use the 'lumped capacity' method of estimating the temperature of a body under unsteady heat flow conditions?

A solid steel sphere of radius 10 mm and a solid steel cylinder of radius 5 mm and length 10 mm, both initially at a temperature of 100 °C, are immersed in a large reservoir of cold water at 20 °C. After 1 minute the sphere is at a temperature of 50 °C. Estimate from basic principles the temperature of the cylinder after 1 minute. The specific heat and density of the steel are 0.5 kJ/kg K and 7000 kg/m³.
(31.5 °C)

18. A long copper rod of 10 mm diameter and initially at a temperature of 100 °C is placed in an air stream which has a temperature, pressure and velocity of 27 °C, 1 bar and 2 m/s. Determine the time taken for the rod to cool to 30 °C. Convection is described by the relationship $Nu = 0.7 Re^{0.5}$ (based on the rod diameter). Copper has a specific heat and density of 0.38 kJ/kg K and 8950 kg/m³ respectively.
(6.9 min)

19. A cup of tea cools by passing most of its heat to the surroundings in three ways: from the tea surface, through the sides of the cup and through the base. A cylindrical cup has a diameter of 70 mm, a depth of 70 mm, a wall thickness of 3 mm and a thermal conductivity of 1 watt/m K. It is full to within 10 mm from the brim with tea (which may be assumed to have the physical properties of water) and the heat transfer coefficient from the cup

side is 0.04 kW/m² K and the top surface is 0.1 kW/m² K. The heat flow through the base is negligible. Determine:

(a) the relationship between the heat flow rate and the temperature difference between the tea and the surroundings, assuming the tea is at a uniform temperature;

(b) the time taken for the tea to cool from 70 °C to 50 °C in surroundings at 20 °C.

(8.8 min)

20. A sphere of material density ρ is placed in surroundings at a high temperature T_f where it is heated to its sublimation temperature T_s. Assuming that it remains spherical as it sublimes derive an expression for the time taken to sublime in terms of the heat transfer coefficient h, the heat of sublimation Δh_s per unit mass, the temperature difference $(T_f - T_s)$ and the initial volume V. It may be assumed that the thermal resistance at the surface of the sphere is much greater than the resistance of the sphere material.

21. A steel tube of length 20 cm with internal and external diameters of 10 cm and 12 cm is quenched from 500 °C to 30 °C in a large reservoir of water at 10 °C. Below 100 °C the heat transfer coefficient is 1.5 kW/m² K. Above 100 °C it is less owing to a film of vapour being produced at the surface, and an effective mean value between 500°C and 100 °C is 0.5 kW/m² K. The density of the steel is 7800 kg/m³ and the specific heat is 0.47 kJ/kg K. Determine the quenching time.
(77 s)

22. A small sphere of radius r and a cylinder of radius r and length r initially both at the same temperature are immersed in a reservoir of cool fluid with density ρ and specific heat c. If the heat transfer coefficient h for each has the same value show that after a time τ the ratio of their temperatures (with respect to the fluid temperature) is given by:

$$\frac{\Delta T_s}{\Delta T_c} = \exp\left(+ \frac{h\tau}{\rho cr} \right)$$

23. Broad beans are sown at a depth of 10 cm in a soil which has a conductivity of 0.6×10^{-3} kW/m K and a thermal diffusivity of 0.1735×10^{-6} m²/s.

During the following night and morning the air temperature varies in the following manner:

Time	Air Temperature (°C)
8 p.m.	2
10 p.m.	−1
12 midnight	−3
2 a.m.	−4
4 a.m.	−4

6 a.m.	-3
8 a.m.	0
10 a.m.	2
12 noon	3

The heat transfer coefficient between the air and the ground is $0.012 \text{ kW/m}^2 \text{ K}$ and at 8 p.m. it may be assumed that the soil is at a steady temperature of $2\,^\circ\text{C}$.

Estimate (a) the time when the beans become frozen at $0\,^\circ\text{C}$.

(b) the duration of time for which they are frozen (to the nearest hour).

(7.0 a.m., 4 h)

24. At a certain time during late evening on the moon a very large slab of moonrock is found to have a uniform temperature of $0\,^\circ\text{C}$. During the lunar night, heat is transferred from the rock to outer space at a temperature of $0\,^\circ\text{K}$. The rock has an emissivity of unity, a thermal conductivity of 2 W/m K and a thermal diffusivity of $2.78 \times 10^{-6} \text{ m}^2/\text{s}$.

Commence a graphical solution to the temperature distribution in the slab using the Schmidt finite difference technique and hence show that after 6 hours the surface temperature of the slab has fallen by about $35\,^\circ\text{C}$. (It will be necessary to calculate the radiation heat transfer coefficient progressively. A scale of 1 cm represents 0.2 m is suggested.)

25. A concrete surface at an initial steady temperature of $50\,^\circ\text{C}$ is cooled by a stream of water at $10\,^\circ\text{C}$. The mean heat transfer coefficient between the water and the concrete is $0.2 \text{ kW/m}^2 \text{ K}$ and the conductivity and thermal diffusivity are 2 W/m K and $1 \times 10^{-6} \text{ m}^2/\text{s}$. Estimate using the graphical technique the time for the concrete at a depth of 2 cm to reach $30\,^\circ\text{C}$.
($13\frac{1}{2}$ min)

26. Write a short account of the mechanisms of thermal energy transfer in liquids and describe an experimental arrangement suitable for the determination of the thermal conductivity of oil.

10.4 Convection and Heat Exchangers (Chapters 3, 5 and 8)

10.4.1 Summary

Surface heat flux

$$q = -h\,\Delta T \tag{3.1}$$

Reynolds analogy (turbulent flow, $\text{Pr} = 1$)

$$q = -\frac{c\tau_0}{U}\Delta T \tag{3.5}$$

Dimensionless representation of forced convection

$$Nu = C\,Re^n\,Pr^m$$

where Nusselt number, $Nu = hd/k$

Reynolds number, $Re = \rho Ud/\mu$

Prandtl number, $Pr = c\mu/k$

and C, n and m are constants.

Dimensionless representation of natural convection

$$Nu = C\,Ra^n$$

where Rayleigh number

$$Ra = \frac{g\beta\,\Delta T d^3 c_p \rho}{\mu k}$$

Heat exchangers
mass continuity

$$\dot{m} = \rho U A_c$$

energy conservation

$$-\dot{m}_h c_h\,\Delta T_h = \dot{m}_c c_c\,\Delta T_c \tag{5.1}$$

where suffixes h and c refer to hot and cold fluids.
Log mean temperature difference ΔT_m

$$Q = UA\,\Delta T_m \tag{5.9}$$

where

$$\Delta T_m = \frac{\Delta T_2 - \Delta T_1}{\ln(\Delta T_2/\Delta T_1)} \tag{5.10}$$

and suffixes 1 and 2 refer to ends of the exchanger.
Effectiveness of a parallel-flow exchanger

$$E_p = \frac{1 - \exp[-NTU(1 + C)]}{1 + C} \tag{5.17}$$

Effectiveness of a counter-flow exchanger

$$E_c = \frac{1 - \exp[-NTU(1 - C)]}{1 - C\exp[-NTU(1 - C)]} \tag{5.18}$$

Boundary layer equations
mass continuity

$$\frac{\partial u}{\partial x} + \frac{\partial v}{\partial y} = 0 \tag{8.1}$$

momentum

$$u\frac{\partial u}{\partial x} + v\frac{\partial u}{\partial y} = v\frac{\partial^2 u}{\partial y^2} \tag{8.2}$$

225

energy

$$u\frac{\partial T}{\partial x} + v\frac{\partial T}{\partial y} = \alpha\frac{\partial^2 T}{\partial y^2} \tag{8.3}$$

heat flow equation

$$\frac{\mathrm{d}}{\mathrm{d}x}\left(\int_0^Y (T_\infty - T)u\,\mathrm{d}y\right) = \alpha\left(\frac{\mathrm{d}T}{\mathrm{d}y}\right)_0 \tag{8.4}$$

Laminar flow in tubes
radial flow energy equation

$$\frac{1}{r}\frac{\partial}{\partial r}\left(r\frac{\partial T}{\partial r}\right) = \frac{u}{\alpha}\frac{\partial T}{\partial x} \tag{8.5}$$

velocity profile

$$u = 2U\left(1 - \frac{r^2}{r_0^2}\right) \tag{8.7}$$

heat transfer—constant wall heat flux

$$\mathrm{Nu} = 4.364 \tag{8.15}$$

heat transfer—constant wall temperature

$$\mathrm{Nu} = 3.658 \tag{8.19}$$

length L of starting section

$$\frac{L}{d} = 0.0575\,\mathrm{Re} \tag{8.20}$$

Turbulent flow in tubes Dittus–Boelter equation

$$\mathrm{Nu} = 0.023\,\mathrm{Re}^{0.8}\,\mathrm{Pr}^n \tag{8.22}$$

where

$$n = 0.4 \text{ for fluid being heated}$$
$$n = 0.3 \text{ for fluid being cooled.}$$

Non-circular tubes; use d_e in place of d

$$d_e = \frac{4A_c}{P} \tag{8.24}$$

where

$$d_e = \text{equivalent diameter}$$
$$P = \text{wetted perimeter}$$
$$A_c = \text{cross-sectional area}$$

Laminar flow over a flat plate

velocity distribution

$$\frac{u}{u_\infty} = \frac{3}{2}\left(\frac{y}{\delta}\right) - \frac{1}{2}\left(\frac{y}{\delta}\right)^3 \tag{8.26}$$

temperature distribution

$$\frac{\Delta T}{\Delta T_\infty} = \frac{3}{2}\left(\frac{y}{\delta_t}\right) - \frac{1}{2}\left(\frac{y}{\delta_t}\right)^3 \tag{8.27}$$

mean heat transfer—constant wall temperature

$$\text{Nu} = 0.664 \text{ Re}^{0.5} \text{ Pr}^{0.33} \tag{8.33}$$

where the physical dimension is the plate length and fluid properties at the arithmetic mean of the wall and free-stream temperature are used.

Turbulent flow over a flat plate—mean heat transfer:

$$\text{Nu} = 0.036 \text{ Re}^{0.8} \text{ Pr}^{0.33} \tag{8.35}$$

(and the above comments apply)

Cross-flow over a cylinder—mean heat transfer:

gases $\qquad\qquad\quad \text{Nu} = C \text{ Re}^n \tag{8.40}$

liquids $\qquad\qquad\quad \text{Nu} = 1.11 \ C \text{ Re}^n \text{ Pr}^{0.33} \tag{8.41}$

where C and n are given in Table 8.1.

Flow over a sphere—mean heat transfer:

gases $\qquad\qquad\quad \text{Nu} = 0.37 \text{ Re}^{0.6} \tag{8.42}$

Tables and charts

Some properties of common fluids	Table 3.1
Convection relationships	Table 3.2
Simplified convection relationships for air	Table 3.3
Dimensionless groups	Table 3.4
The magnitude of h	Table 3.5
The variation of friction coefficients with Re	Figure 3.4
Summary of log mean temperature difference analysis of exchangers	Figure 5.3
Cross-flow over a cylinder—correlation constants	Table 8.1

10.4.2 Symbols and units (Chapters 3, 5 and 8)

A $\qquad$ Surface area (m^2)

A_c $\qquad$ Cross-sectional area (m^2)

C	Capacity rate ratio for heat exchangers
C_d	Average drag coefficient
c	Specific heat of liquid or gas (at constant pressure) (kJ/kg K)
c_c	Specific heat of cool fluid—exchangers (kJ/kg K)
c_h	Specific heat of hot fluid—exchangers (kJ/kg K)
c_p	Specific heat at constant pressure (kJ/kg K)
d	Dimension or diameter (m)
d_e	Equivalent diameter, equation 8.24 (m)
E	Effectiveness
E_c	Effectiveness of counter-flow exchanger
E_p	Effectiveness of parallel-flow exchanger
F	Correction factor—exchangers
f	Friction factor (Fanning)
G	Fluid mass flow rate per unit area (kg/s m^2)
Gr	Grashof number—natural convection
g	Gravitational constant (m/s^2)
H	Enthalpy (kJ)
h	Mean heat transfer coefficient (kW/m^2 K)
h_{fg}	Specific enthalpy of vaporization (kJ/kg)
h_i	Mean heat transfer coefficient for inner surface of tube (kW/m^2 K)
h_o	Mean heat transfer coefficient for outer surfaces of tube (kW/m^2 K)
h_x	Local heat transfer coefficient (kW/m^2 K)
k	Thermal conductivity (kw/m K)
L	Starting length to fully-developed flow profile (m)
l	Length (m)
$\dot{m}$	Mass flow rate (kg/s)
NTU	Number of transfer units—exchangers
Nu	Nusselt number (mean)
Nu$_x$	Nusselt number at distance x from leading edge of plate
P	Wetted perimeter (m)
Pr	Prandtl number
p	Pressure (bar or N/m^2)
Q	Heat transfer rate (kW)
q	Heat transfer rate per unit area (kW/m^2)
Ra	Rayleigh number—natural convection
Re	Reynolds number (mean)
Re$_{crit}$	Critical Reynolds number (at transition from laminar to turbulent flow)
Re$_x$	Reynolds number at distance x from leading edge of plate
r	Radius (m)
r_0	Radius to wall of tube (m)
T	Temperature (°C or °K)
T_f	Arithmetic mean temperature between wall and fluid
T_0	Temperature at wall (°C)

T_m	Bulk fluid temperature—equation (5.2) (°C)
T_∞	Temperature in bulk of fluid outside the boundary layer (°C)
ΔT	Temperature difference (°C)
ΔT_c	Temperature difference of cool fluid (°C)
ΔT_h	Temperature difference of hot fluid (°C)
ΔT_m	Log mean temperature difference (°C)
ΔT_1	Temperature difference between fluids on left-hand side of exchanger (°C)
ΔT_2	Temperature difference between fluids on right-hand side of exchanger (°C)
ΔT_∞	Temperature difference between bulk of fluid and wall (°C)
t	Time (s)
U	Bulk velocity or free-stream velocity (m/s)
U	Overall heat transfer coefficient (kW/m² K)
u	Velocity, generally in x direction or along tube (m/s)
u_0	Velocity in centre of tube (m/s)
u_∞	Velocity of free stream outside boundary layer (m/s)
V	Velocity of fluid (m/s)
v	Specific volume, i.e. $1/\rho$ (m³/kg)
v	Velocity in y direction (m/s)
v_r	Velocity in radial direction (m/s)
w	Velocity in z direction (m/s)
x	Distance from leading edge of flat plate (m)
y	Distance from wall (m)
α	Alpha—Thermal diffusivity (m²/s)
β	Beta—Coefficient of volumetric expansion (K⁻¹)
δ	Delta—Momentum boundary layer thickness (m)
δ_t	Delta—Thermal boundary layer thickness (m)
ε	Epsilon—Eddy diffusivity (m²/s)
ζ	Zeta—Boundary layer thickness ratio δ_t/δ
θ	Theta—Non-dimensional temperature difference
μ	Mu—Dynamic viscosity (kg/m s)
μ_0	Mu—Dynamic viscosity of fluid at the wall temperature (kg/m s)
v	Nu—Kinematic viscosity, i.e. μ/ρ (m²/s)
ρ	Rho—Density (kg/m³)
τ	Tau—Fluid shear stress (N/m²)
τ_0	Tau—Fluid shear stress at the wall (N/m²)

10.4.3 Problems

1. A condenser contains 100 tubes of 20 mm nominal diameter and 2 m length on which saturated steam condenses at a pressure of 0.5 bar. Cooling water enters the tubes at 20 °C, leaves at 40 °C and flows at a velocity of

0.5 m/s. The steam side condensing heat transfer coefficient is 10 kW/m² K. Estimate the condensate flow rate and the cooling water side heat transfer coefficient.

Compare the steam side and cooling water side heat transfer coefficients and sketch a suitable design of tube which (for the same nominal diameter) would increase the condensate flow rate by a factor of about two.
($h_w = 2.6$ kW/m² K)

2.   Show how Reynolds analogy leads to a relationship between the rate of heat transfer from a fluid in turbulent motion to a wall and the shear stress exerted on the wall.

The internal cross-section of a pipe is an equilateral triangle of side 25 mm. The pressure drop per metre along the pipe is 0.2×10^{-3} bar at a gas velocity through the pipe of 20 m/s. The mean constant pressure specific heat of the gas is 1.1 kJ/kg K. Estimate the mean heat transfer coefficient between the gas and the pipe wall.
(4.0 W/m² K)

3.   Reynolds analogy of turbulent forced convection heat transfer leads to a relationship between the heat flow rate q and the fluid shear stress τ. Derive this relationship and show that it is modified by the term in brackets:

$$\frac{q}{\tau_a} = \frac{c_p \, \Delta T}{U} \left[\frac{1}{1 - a(\mathrm{Pr} - 1)} \right]$$

when account is taken of a laminar sub-layer (with a velocity at the laminar–turbulent transition a fraction 'a' of the bulk velocity). Derive a relationship between the heat transfer coefficient and the sub-layer thickness in terms of the fluid properties and the fraction 'a'.

4.   The hotter fluid in a parallel-flow heat exchanger decreases in temperature from 200 °C to 100 °C and the cooler fluid increases from 40 to 90 °C. For the same heat transfer what is the percentage saving in area made by using counter-flow instead of parallel-flow? What are the temperatures of the two fluids at the mid-point of the exchanger in each case?
(34.5%; 60 °C, 141.5 °C)

5.   (a) Show for steam condensing in the annulus of a double tube heat exchanger that the mid-point temperature is the same whether the coolant is flowing in 'parallel'- or 'counter'-flow.

(b) Give an expression for the log mean temperature difference in the case of a counter-flow double tube heat exchanger when the product $\dot{m}c$ for both fluids has the same value.

6.   Engine oil circulates at the rate of 500 kg/h and is cooled by water from 100 °C to 50 °C in a single pass counter-flow heat exchanger. The water enters at 20 °C and leaves at 80 °C. The overall heat transfer rate is typically 1 kW/m² K for this type of exchanger and the specific heat of the oil is

1.8 kJ/kg K. Determine the water flow rate and the effectiveness of the exchanger. Use may be made of the following equation if desired:

$$E = \frac{1 - \exp[-NTU(1 - C)]}{1 - C\exp[-NTU(1 - C)]}$$

where NTU is the number of transfer units and C is the capacity rate ratio. (179 kg/h, 75%)

7. A condenser contains 100 thin-walled tubes of 25 mm nominal diameter and 2 m length. Cooling water enters each tube at a temperature of 10 °C, leaves at 60 °C and flows at a velocity of 2 m/s. The heat transfer coefficient between the cooling water and the tube may be calculated using the Dittus–Boelter relationship:

$$Nu = 0.023\ Re^{0.8}\ Pr^{0.4}$$

together with property values from tables. The condensing heat transfer coefficient may be taken as 5 kW/m² K and the condensate temperature as 80 °C. Estimate the rate at which the steam is condensed.
(0.86 kg/s)

8. An initial feasibility study of a Rankine cycle steam power plant suitable for a large lorry involves estimation of the number of air cooled condenser tubes required. Steam is condensed within tubes which are 300 mm long and of 50 mm internal diameter and the fin geometry is such that the effective air side area of the tubes is 20 times the steam side. The inner and outer heat transfer coefficients for this arrangement are 5 and 0.05 kW/m² K respectively. Estimate the number of tubes required under conditions such that the air enters at 25 °C and leaves at 55 °C, and 0.2 kg/s of saturated steam at 10 bar are condensed.
(73)

9. Discuss the influence of filmwise and dropwise condensation and the effect of air on the heat transfer rate in a steam condenser.

A steam condenser operates at 0.5 bar and contains 100 tubes each of 2 m length and 25 mm nominal diameter. Cooling water enters the tubes at 20 °C, leaves at 50 °C and has a heat transfer coefficient to the tube wall of 1 kW/m² K. Under normal filmwise condensation the condensate flow rate is found to be 0.25 kg/s. Estimate the condensate flow rate under dropwise conditions if the steam side heat transfer coefficient is increased by a factor of 10 over the filmwise heat transfer coefficient. (It may be assumed that the cooling water flow rate is adjusted so that the inlet and exit temperatures are unchanged.)
(0.3 kg/s)

10. In what situations does the 'effectiveness' approach to heat exchanger calculations have advantages over the 'log mean temperature difference' approach?

In a tubular counter-flow heat exchanger 0.3 kg/s of water are heated from 40 to 80 °C by hot gases ($c_p = 1.0$) which enter at 200 °C and leave at 100 °C. The overall heat transfer coefficient is 0.2 kW/m² K. Calculate the area of the heat exchanger using:

(a) the log mean temperature difference approach; and

(b) the effectiveness—*NTU*—approach

(For a counter-flow heat exchanger the effectiveness is given by:

$$E = \frac{1 - \exp[-NTU(1 - C)]}{1 - C\exp[-NTU(1 - C)]}$$

where C is the capacity rate ratio.)
(2.88 m²)

11. Show from basic principles that the effectiveness E of a parallel-flow double-pipe heat exchanger, in which the temperature change of one fluid is very much less than the temperature change of the other fluid, is given approximately by:

$$E = 1 - e^{-n}$$

where n denotes the number of transfer units.

12. Oil with a mean specific heat of 2.5 kJ/kg K is to be cooled from 110 °C to 30 °C in a single-pass counter-flow heat exchanger. The coolant is water which enters at 20 °C and leaves at 80 °C and the overall heat transfer coefficient for this type of exchanger is 1.5 kW/m² K.

If the water flow rate is 1500 kg/h determine the quantity of oil that can be cooled per hour and the heat exchanger area. What are the fluid exit temperatures when the water flow rate is decreased to 1000 kg/h for the same oil flow rate? The effectiveness of a counter-flow heat exchanger is given by the following expression where *NTU* is the number of transfer units and C is the capacity rate ratio:

$$E = \frac{1 - \exp[-NTU(1 - C)]}{1 - C\exp[-NTU(1 - C)]}$$

(40.3 °C, 98.6 °C)

13. Milk is passed through a cooler at the rate of 5 litre/minute and is reduced in temperature from 36 °C to 20 °C. The cooler consists of six metal tubes of 10 mm internal diameter and 2 m long with milk flowing in parallel through the tubes and water flowing around the outside of the tubes. The water maintains the tube temperature at approximately 15 °C throughout.

Estimate the tube to milk heat transfer coefficient and check your estimation by application of the Dittus–Boelter relationship:

$$Nu = 0.023\ Re^{0.8}\ Pr^{0.3}$$

It may be assumed that milk has the properties of water.

232

14. Explain why the forced convection heat transfer coefficient increases with increase in fluid velocity in the case of turbulent flow but not in the case of laminar flow.

Air is passed through a heater into a room which is maintained at 20 °C. The heater consists of tubes which are maintained at 100 °C by condensing steam. A third of the air entering the heater comes from the room and the other two-thirds from the surroundings at 10 °C. To maintain these conditions it is found that the mass flow rate through the heater is 0.2 kg/s. When the temperature of the surroundings drops to 5 °C what is the new mass flow rate to maintain the same room temperature? It may be assumed that the air is dry and the flow is turbulent such that $Nu \propto Re^{0.7}$. Ignore the variation of air properties with temperature.
(0.057 kg/s)

15. Explain the differences between the thermal diffusivity, momentum diffusivity (or kinetic viscosity) and the eddy diffusivity. By consideration of the diffusion of momentum and heat from a surface to a surrounding turbulent fluid under forced convection conditions, derive an approximate relationship between the heat flow rate and the wall shear stress.

Use this relationship to estimate the mean heat transfer coefficient between the wall of a square section tube of side 25 mm and air at a temperature of 100 °C flowing at the rate of 0.006 kg/s with a pressure loss per metre length of 10 N/m². Check that the flow is turbulent in nature. Why is the derived relationship less applicable to water than to air?
(6.2 W/m² K)

16. (a) By considering the energy transfer to a differential control volume within a boundary layer on a flat plate and making reasonable assumptions, derive the energy equation in the following form:

$$u \frac{\partial T}{\partial x} + v \frac{\partial T}{\partial y} = \alpha \frac{\partial^2 T}{\partial y^2}$$

where u is the velocity in the x direction parallel to the plate, v is the velocity in the y direction and α is the thermal diffusivity. Explain why this equation is not applicable when the velocity of the free steam is extremely low or extremely high.

(b) By expressing T (or preferably ΔT) as a polynominal in y, show that the temperature distribution in the layer thickness $y = \delta_t$, may be expressed in the form:

$$\frac{T - T_0}{T_\infty - T_0} = \frac{\Delta T}{\Delta T_\infty} = \frac{3}{2}\left(\frac{y}{\delta_t}\right) - \frac{1}{2}\left(\frac{y}{\delta_t}\right)^3$$

where suffixes '0' and '∞' indicate the plate surface and free stream respectively.

17. Forced convection in fully-developed laminar flow through a tube may be analysed by assuming that heat flow in the radial direction is entirely by conduction. Thus q the radial heat flow per unit length of tube of radius r is given by:

$$q = -k2\pi(r - y)\frac{d\theta}{dy}$$

where θ is the temperature difference between the wall and an annulus dy at a distance y from the wall. By expressing this temperature difference as $\theta = ay + by^2 + cy^3$ and evaluating the constants show that the Nusselt number (based on the tube diameter and the temperature difference between the tube centre and the wall) is equal to 2.4.

The fully-developed laminar flow of water along a 20 mm diameter tube is found to yield a mean heat transfer coefficient of 0.124 kW/m² K.

Give the reason for the considerable difference between the experimental Nusselt number and that obtained using the above theory.

18. Reynolds analogy of turbulent-flow convective heat transfer over a flat plate leads to the following relationship between the dimensionless parameters

$$Nu_x = \frac{f}{2} Re\,Pr$$

where f is the friction coefficient and where the suffix x denotes that the Nusselt number is the local value at distance x from the leading edge. By substitution of the turbulent flow relationship $f = 0.059\,Re^{-0.2}$ derive an expression for the mean Nusselt number Nu. Comment on the heat transfer predictions of this expression compared to the predictions of the following empirical relationship for the cases of air and crude oil.

$$Nu = 0.036\,Re^{0.8}\,Pr^{0.33}$$

Water flows over a flat plate measuring 1 m × 1 m at the rate of 2 m/s. The plate is at a uniform temperature of 90 °C and the water temperature is 10 °C. The heat transfer relationship for laminar flow is

$$Nu = 0.664\,Re^{0.5}\,Pr^{0.33}$$

and the critical Reynolds number is 0.5×10^6. Estimate the length of plate over which the flow is laminar and the heat flow rate to the entire plate. (0.138 m, 434 kW)

19. It is desired to measure the temperature of a jet of hot air using a mercury in glass thermometer. The thermometer is 5 mm diameter and 250 mm of its length protrudes into the hot air stream. The emissivity may be taken as 0.1 and the temperature of the surroundings is 20 °C. The air stream

has a mean velocity of 20 m/s and a nominal temperature of 200 °C. Heat transfer by forced convection to a cylinder under these conditions is given by the expression

$$Nu_d = 0.17 \, Re_d^{0.62}$$

and heating caused by air friction on the thermometer may be taken as 0.1 milliwatt. By carrying out a heat balance on the thermometer determine the difference between the indicated temperature and the air stream temperature.

(11 °C)

20. Analyse film condensation on a flat vertical plate by considering the shear, gravity and vapour forces acting on the condensate layer. Derive an expression for the condensate velocity and by integrating across the layer thickness δ determine the mass flow rate. Show that if the vapour density is much less than the liquid density the Reynolds number may be given by:

$$Re = \frac{4}{3} \frac{g\rho^2\delta^3}{\mu^2}$$

The symbols have the usual connotation.

Conversion Factors to British Units

Mass	m	1 kg	= 2.2046 lb
Length	l	1 m	= 3.2808 ft
Time	t	1 s	= 2.778 × 10⁻⁴ h
Temperature	T	1 K	= 1.8 °R
		x °C	= $(1.8x + 32)$ °F
Force		1 N (Newton)	= 0.2248 lbf
Density	ρ	1 kg/m³	= 0.06243 lb/ft³
Pressure	p	1 N/m²	= 1 Pascal
		1 bar	= 10⁵ N/m² = 14.5 lbf/in²
Energy		1 kJ	= 0.9478 Btu
Energy flow rate (Power)	Q	1 kJ/s (1 kW)	= 3412 Btu/h = 1.341 hp
Energy flow rate per unit area	q	1 kW/m²	= 317.0 Btu/ft² h
Specific heat	c	1 kJ/kg K	= 0.2388 Btu/lb °F
Thermal conductivity	k	1 kW/m K	= 577.7 Btu/ft h °F
Thermal diffusivity	α	1 m²/s	= 38,740 ft²/h
Dynamic viscosity	μ	1 kg/m s	= 2419 lb/ft h
			= 10 poise
Heat transfer coefficient	h	1 kW/m² K	= 176.1 Btu/ft² h °F

Approximate Material Properties (at 1 Atmosphere)

Material	T °C	ρ kg/m³	k W/m K	c_p kJ/kg K	μ × 10⁶ kg/m s
Fluids					
Air	20	1.21	0.0257	1.005	18.1
	100	0.94	0.032	1.01	2.18
Hydrogen	20	0.084	0.178	14.3	8.83
	100	0.063	0.216	14.4	10.4
Steam	100	0.60	0.025	1.9	12.2
Water	20	1000	0.60	4.18	1002
	100	960	0.68	4.22	279
Freon-12 (liquid)	20	1330	0.073	0.97	270
Crude oil	20	∼900	∼0.1	∼2.0	∼50,000
Mercury	20	13,500	8.0	0.14	1570

Material	T °C	ρ kg/m³	k W/m K	c_p kJ/kg K	ε (emissivity)
Solids					
Copper	20	8950	385	0.383	0.03 polished
Aluminium	20	2710	204	0.896	0.2 oxidized
Mild steel	20	7830	54	0.465	0.8 rusted
Stainless steel (18/8)	20	7820	16	0.460	0.07 polished
Granite	20	2640	3.0	0.82	—
Ice	0	913	2.2	1.93	0.92 (sheet)
Concrete	20	2200	1.4	0.88	0.8
Wood (pine, across grain)	20	700	1.4	2.6	0.9
Glass	20	2700	0.8	0.84	0.94
Common brick	20	1600	0.7	0.84	0.93
Bakelite	20	1280	0.23	1.6	—
Asbestos	20	580	0.16	0.82	0.96
	100	580	0.19	0.82	—
Glass wool	20	200	0.04	0.70	—
Cork (expanded)	20	100	0.036	1.9	0.95
Polyurethane foam	20	80	0.03	—	—

References

ADAMS, J. A. and ROGERS, D. E. (1973) *Computer Aided Heat Transfer Analysis*, McGraw-Hill, New York.

ALLEN, D. N. de G. (1954) *Relaxation Methods*, McGraw-Hill, New York.

ANON (1971) *Modern Power Station Practice*, CEGB, Pergamon, Oxford.

ASHRAE (1967) Am. Soc. Heating, Refrigeration and Air Conditioning Engineers, *Handbook of Fundamentals*.

BANKOFF, S. G. (1958) *A.I.Ch.E.J.*, **4**, 24.

BIOT, J. B. (1804) *Bibliothéque Britannique*, **27**, 310.

BOLTZMANN, L. (1884) *Wiedemanns Annalen*, **22**, 291.

BOTTERILL, J. S. M. (1974) *Fluidized Bed Heat Transfer*, Academic, New York.

BRIDGEMAN, P. W. (1923) *Proc. Nat. Acad. Sciences*, **9**, 341.

CARSLAW, H. S. and JAEGER, J. C. (1959) *Conduction of Heat in Solids*, Oxford, New York.

CHAPMAN, A. J. (1974) *Heat Transfer*, Macmillan, New York.

CHISHOLM, D. (1971) *The Heat Pipe*, Mills and Boon, London.

COLBURN, A. P. (1933) *Trans. A.I.Ch.E.*, **29**, 174.

COLE, R. (1974) *Adv. in Heat Transfer*, **10**, 86.

COLLIER, J. G. (1972) *Convective Boiling and Condensation*, McGraw-Hill, New York.

CORNWELL, K. (1971) *J. Phys., D., Appl. Phys.*, **4**, 441.

CORNWELL, K. (1975) *A.I.Ch.E. Annual Meeting*, Paper **114**, Los Angeles.

DAVIDSON, J. F. and HARRISON, D. (1971) *Fluidization*, Academic Press, London.

DESCHANEL, A. P. (1888) *Natural Philosophy, Part 2, (Heat)*, Blackie, London.

DIAMONT, R. M. E. (1964) *Heating and Ventilating Engineer, Parts 1–7*, January–July.

DITTUS, F. W. and BOELTER, L. M. K. (1930) *Univ. California Pub. Engng.*, **2**, 443.

DUNKLE, R. V. (1954) *Trans. A.S.M.E.*, **76**, 549.

DUSINBERRE, G. M. (1961) *Heat Transfer Calculation by Finite Differences*, Int. Textbook Co.

ECKERT, E. R. G. and DRAKE, R. M. Jr. (1972) *Analysis of Heat and Mass Transfer*, McGraw-Hill, New York.

ECKERT, E. R. G. and GROSS, J. F. (1963) *Introduction to Heat and.Mass Transfer*, McGraw-Hill, New York.
ECKERT, E. R. G. and JACKSON, T. W. (1951) *N.A.C.A. Rept.*, **1015**.
ELDER, J. W. (1965) *J. Fluid Mech.*, **23**, 77.
FANGER, P. O. (1970) *Thermal Comfort*, McGraw-Hill, New York.
FINLAY, I. C. (1975) *Chartered Mech. Eng'r.*, **22**, 3, 59.
FORSTER, H. K. and ZUBER, N. (1955) *A.I.Ch.E.J.*, **1**, 535.
FOURIER, J. B. J. (1822) Théorie Analytique de la Chaleur, Paris.
GAERTNER, R. F. (1965) *J. Heat Transfer*, **87**, 17.
GOLDSTEIN, R. J. and CHU, T. Y. (1969) *Progr. Heat Mass Transfer*, **2**, 55.
GRIFFITH, P. and WALLIS, J. D. (1960) *Chem. Eng. Progr. Symp. Ser.*, **56**, 49.
HAAS, W. J. de and BIERMASZ, Th. (1935) *Physica*, **2**, 673.
HAMILTON, D. C. and MORGAN, W. R. (1952) *NACA, TN 2836*.
HEISLER, M. P. (1947) *Trans. A.S.M.E.*, **69**, 227.
HILPERT, R. (1933) *Forsch. Gebiete Ingenieurw.*, **4**, 215.
HOLMAN, J. P. (1966) *Heat Transfer*, McGraw-Hill, New York.
HOTTEL, H. C. and SAROFIM, A. F. (1967) *Radiative Transfer*, McGraw-Hill, New York.
INGENHOUSZ, I. (1789) *Journ. de Physique*, **34**, 68 and 380.
JAKOB, M. (1949, 1957) *Heat Transfer*, Vols I and II, Wiley, New York.
KARMAN, Th. Von. (1939) *Trans. A.S.M.E.*, **61**, 705.
KAYS, W. M. (1966) *Convective Heat and Mass Transfer*, McGraw-Hill, New York.
KAYS, W. M. and LONDON, A. L. (1964) *Compact Heat Exchangers*, McGraw-Hill, New York.
KERN, D. Q. (1950) *Process Heat Transfer*, McGraw-Hill, New York.
KIRCHHOFF, G. (1859) *Monatsber d. preuss. Akad. d. Wiss.*, p. 783.
KITTEL, C. (1956) *Introduction to Solid State Physics*, Wiley, New York.
KNUDSEN, J. D. and KATZ, D. L. (1958) *Fluid Dynamics and Heat Transfer*, McGraw-Hill, New York.
KRAUSSOLD, H. (1936) *Fors. Geb. Ingen.*, **2**, 186.
KUNII, D. and LEVENSPIEL, O. (1969) *Fluidization Engineering*, Wiley, New York.
LANGHAAR, H. L. (1942) *J. Appl. Mech.*, **64**, A55.
LEPPERT, G. and PITTS, C. C. (1964) *Adv. in Heat Transfer*, **1**, 185.
LEVENSPIEL, O. and WALTON, J. S. (1954) *Chem. Eng. Progr. Symp. Ser.*, **50**, 9, 1.
LINDSAY, R. *et al.* (1970) *British Patent No. 1338767*.
LORENZ, L. (1872) *Poggendorffs Annalen*, **147**, 429.
MALAVARD, L. C. (1956) *The Use of Rheoelectrical Analogies in Aerodynamics, AGARD-ograph 18, NATO*.
MAYHEW, Y. R. and ROGERS, G. F. C. (1969) *Thermodynamic and Transport Properties of Fluids*, Oxford, London.
MCADAMS, W. H. (1954) *Heat Transmission*, McGraw-Hill, New York.
MIKIC, B. B. and ROHSENOW, W. M. (1969) *J. Heat Transfer*, **91**, 245.
MYERS, G. E. (1971) *Analytical Methods in Conduction Heat Transfer*, McGraw-Hill, New York.
NEWTON, I. (1701) *Phil. Trans. Roy. Soc., London*, **22**, 824.
NUKIYAMA, S. (1934) *J. Soc. Mech. Engrs. Japan*, **37**, 367.
NUSSELT, W. (1916) *Die Ober, des Wass., V.D.I. Zeit.*, **60**, 541 and 569.
OSTRACH, S. (1952) *N.A.C.A., Tech. Note 2635*.
PEIERLS, R. A. (1955) *Quantum Theory of Solids*, Oxford, London.
PLANCK, M. (1901) *Annalen. d. Physik*, **4**, 553.
POOLE, M. J. (1967) *Brit. Nuclear Energy Soc. J.*, **6**, 206.
PRANDTL, L. (1910) *Physik. Zeitschr.*, **11**, 1072.

239

References

PRANDTL, L. (1928) *Physik. Zeitschr.*, **29**, 487.

REYNOLDS, O. A. (1874) *Proc. Manchester Lit. and Phil. Soc.*, **14**, 9.

ROHSENOW, W. M. (1952) *Trans. A.S.M.E.*, **74**, 969.

ROHSENOW, W. M. (1956) *J. Heat Transfer*, **78**, 1645.

ROHSENOW, W. M. and GRIFFITH, P. (1956) *Chem. Eng. Progr. Symp. Ser.*, **52**, 47.

SCHLICHTING, H. (1960) *Boundary Layer Theory*, McGraw-Hill, New York.

SCHMIDT, E. (1936) *Einfuhrung in die Technische Thermodynamik*, Springer-Verlag, Berlin.

SCHNEIDER, P. J. (1955) *Conduction Heat Transter*, Addison-Wesley, Reading, Mass.

SHOUKRI, M. and JUDD, R. L. (1975) *J. Heat Transfer*, **97**, 93.

SIEDER, E. N. and TATE, G. E. (1936) *Ind. Engng. Chem.*, **28**, 1429.

SILVER, R. S. (1963) *Proc. I. Mech.E.*, **178**, 339.

SOUTHWELL, R. V. (1940) *Relaxation Methods in Engineering Science*, Oxford, New Jersey.

SPALDING, D. B. and PATANKAR, S. V. (1967) *Heat and Mass Transfer in Boundary Layers*, Morgan-Grampian, London.

SPARROW, E. M. and CESS, R. D. (1966) *Radiation Heat Transfer*, Brooks-Cole, Belmont, Calif.

STEFAN, J. (1879) *Sitzungsber. d. Kais. Akad. d. Wiss. Wien.*, **68**, 385.

TAYLOR, G. I. (1916) *Brit. Adis. Com. Aero. Rep.*, **272**.

TONG, L. S. (1965) *Boiling Heat Transfer and Two-Phase Flow*, Wiley, New York.

TSEDERBERG, N. V. (1965) *Thermal Conductivity of Gases and Liquids*, Arnold, London.

VACHON, *et al.* (1968) *J. Heat Transfer*, **90**, 239.

VITOVITCH, D. and OLSEN, G. H. (1964) *Int. J. Elect. Eng. Educ.*, **1**, No. 3.

VREEDENBERG, H. A. (1958) *Chem. Eng. Sci.*, **9**, 52; (1960), **11**, 274.

WELTY, J. R. (1974) *Engineering Heat Transfer*, Wiley, New York.

WIEN, W. (1893) *Berichte d. preuss. Akad. d. Wiss.*, p. 55.

WIEM, W. (1896) *Annalen d. Physik*, **58**, 662.

WINTER, E. R. F. and BARSCH, W. O. (1971) 'The Heat Pipe', *Adv. in Heat Transfer*, Vol. 8.

ZIMAN, J. M. (1960), *Electrons and Phonons*, Oxford, London.

ZUBER, N. (1958), *J. Heat Transfer*, **80**, 711.

Index